Patrick Moore's Practical A

For other titles published in this series, go to
http://www.springer.com/series/3192

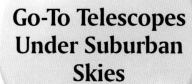

Go-To Telescopes Under Suburban Skies

Neale Monks

 Springer

Neale Monks
HP4 3EH Berkhamsted
Hertfordshire
United Kingdom

ISSN 1431-9756
ISBN 978-1-4419-6850-0   e ISBN 978-1-4419-6851-7
DOI 10.1007/978-1-4419-6851-7
Springer New York Dordrecht Heidelberg London

Library of Congress Control Number: 2010932673

Printed on acid-free paper

Springer is part of Springer Science+Business Media (www.springer.com)

Preface

This book is a compendium of targets for owners of go-to telescopes observing from suburban and exurban (even further out than suburban) locations. Unlike most other books about the deep sky, this book doesn't contain sky charts or star hops, since those aren't needed by owners of go-to telescopes. Instead the aim is to provide information on as many interesting stars, star clusters, nebulae, and galaxies as possible. All are visible from suburban or exurban locations, and none require an aperture greater than 200 mm.

Within each season there are three categories of deep sky objects, and one category for stars. The section on wintertime objects, for example, contains 77 deep sky objects and 28 stars. The first of the deep sky object categories is one for showpiece objects; then there's a category for interesting rather than exceptional objects, and finally a category of objects that are obscure or difficult to see under suburban conditions.

The stars included in each chapter include a variety of double stars, variable stars, unusually colorful stars, and stars that feature in unusual asterisms or clusters. Several stars are included because of their historical or scientific importance. Unlike deep sky objects, stars are largely unaffected by light pollution. This makes them especially rewarding targets for suburban observers.

Although written for owners of go-to telescopes, there's nothing to stop owners of non-computerized telescopes from using this book. Used alongside a star chart or planetarium program, this book could help owners of traditional telescopes get ideas about what's worth observing on a particular night.

One key difference between this book and most other deep sky books is the assumption that the observer will be working under light-polluted skies and using a telescope with an aperture of 200 mm or less. For the purposes of this book, a

200 mm telescope is considered a 'large' telescope, one around 150 mm a 'medium' telescope, and anything less than 100 mm a 'small' telescope.

Indeed, most comments on the brightness of objects will be subjective ones related to aperture and light pollution. Very little will be said about visual magnitudes, since with deep sky objects these values are often very misleading. Instead the reader will be told about how bright the object seems, how much contrast there is between the background sky and the object itself, and whether light pollution filters help to make the object easier to see.

Light pollution is one of the two most limiting issues that affect suburban astronomy (the other being obscuring objects such as trees and buildings close to the horizon). Most of the author's observing was done in three different places, all subject to light pollution of varying severities. My observing in England is done in Berkhamsted, a little over 30 miles from London, and with skies that rate about 6 on the Bortle scale of sky darkness. In the United States, my observing from suburban Lincoln, Nebraska, is under skies of similar quality. Exurban Lincoln is quite a bit better, with the skies at the Olive Creek Recreation Park about 10 miles southwest of the city between 3 and 4 on the Bortle scale.

Although this is primarily a book about northern hemisphere observing, some southern sky objects are included. In my case, these objects were mostly observed during vacations to Hutchinson Island, not far from Stuart, Florida, and at a latitude of 27° north. While many of these objects can still be seen as far north as the American Midwest, observers as far north as southern England will find southern sky targets difficult or impossible to see. Where latitude is relevant to observing an object, it is mentioned in the text, usually with some indication of how far an object rises above the southern horizon, if it does so at all.

This isn't a book about hardware. But that said, two pieces of hardware are so useful that suburban astronomers should consider owning them. The first is a light pollution filter. There are various kinds, each with its own strengths and weaknesses. The second must-have item is a reducer-corrector, a lens that allows Schmidt–Cassegrain telescopes (SCTs) to behave almost like wide field telescopes. More will be said about both of these accessories in the first chapter.

Finally, some words of thanks. The eyepiece simulations used here to suggest what would be seen through a telescope were put together using Starry Night Pro Plus, courtesy of Simulation Curriculum Corp, and AllSky data, courtesy of Main-Sequence Software Inc. The author must thank Pedro Braganca and Doug George for making these excellent tools available to him. The author also wishes to thank Michelle Meskill and Kevin Kawai at Celestron for providing him with photos of Celestron hardware and offering useful comments on the text, particularly with regard to the use and maintenance of go-to telescopes. More valuable comments on the text came from David W. Knisely at the Prairie Astronomy Club in Lincoln, Nebraska. His comments on the benefits of light pollution filters were especially useful. Finally, the help that John Watson and Maury Solomon provided getting this project off the ground cannot be overstated. To all of them, thank you for your help, patience and support.

Two amateur astronomers passed away while I was putting together this book, David Brokofsky (of Lincoln) and David Schultz (of Omaha). In different ways, they each helped me enjoy this hobby and develop my observing skills. To both of them: Clear skies!

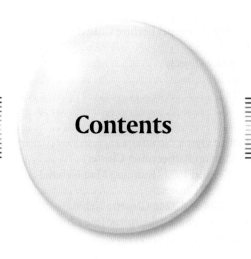

Contents

About the Author

Neale Monks is a scientist, writer, and teacher, and the author of another book in the Practical Astronomy Series, *Astronomy with a Home Computer*. After completing his zoology degree at Aberdeen University he worked briefly as a marine zoologist before moving to London, where he earned his Ph.D. while working at the Natural History Museum. He then spent a few years as a post-doctoral researcher studying the effects of astronomical events on sea level and mass extinctions before leaving research to spend more time teaching and writing. Since 2002 he has taught a history of science class for Pepperdine University as well as various science classes for the WEA. At different times he's lived in England, Scotland, and the Midwestern United States.

CHAPTER ONE

Introduction

How to Use This Book

The four chapters after this one contain objects best placed in winter, spring, summer, or autumn, respectively. So to start with, you'll want to open the book to the chapter relevant to whatever month you're observing in.

Within each chapter, there are four sections. The first section contains those showpiece deep sky objects that you'll want to look at every chance you get. These are often the brightest objects as well and so tend to be the ones most tolerant of light pollution. In some cases, light pollution, narrowband, or line filters may well improve the view, but you should be able to see these objects even without a filter.

The second section covers those deep sky objects that might not be showpiece quality, but are still interesting and worthwhile targets you'll want to include on your observing list. Although these objects are usually visible from suburban locations, a filter of some type may be necessary for some.

The third section includes deep sky objects that are, for various reasons, not often observed. In many cases this is because they are faint, and exurban rather than suburban conditions will probably be necessary (though narrowband or line filters may make it possible to see some of these objects under suburban conditions). Other objects listed in this section would be easy targets for southern hemisphere observers, but for those in the northern hemisphere they are much more difficult. In particular, obstructions such as buildings, trees, and garden fences may make it impossible to see such objects clearly.

N. Monks, *Go-To Telescopes Under Suburban Skies*, Patrick Moore's Practical Astronomy Series, DOI 10.1007/978-1-4419-6851-7_1, © Springer Science+Business Media, LLC 2010

The fourth and final section contains a variety of interesting stars. Some of these will be prettily colored double stars, others notable variable stars, and yet others that have some scientific or cultural significance not obvious at first glance.

One problem with many books about the deep sky is that they use photographs of deep sky objects that, while beautiful, are totally misleading in terms of what you can see through a telescope. Instead this book features simulations indicative of what you're likely to see through a 200 mm aperture telescope under fairly good suburban skies with only low to moderate amounts of light pollution.

Besides the aperture of the telescope, the focal ratio (f-number) of the telescope is also given. In most cases this will be f/10 where the image simulates the telescope operating at its default focal length of 200 mm. But in some cases it will be f/6.3 if the use of a reducer-corrector is being simulated, or else f/20 if a Barlow lens is being simulated.

Also provided is the focal length and type of eyepiece being simulated, either a 35 mm-Panoptic, a 32 mm-Plössl, a 20 mm-Plössl, or a 7 mm-Nagler. In order, these represent a wide field 2-in. eyepiece, a low power 1.25-in. eyepiece, a medium power 1.25-in. eyepiece, and a high power 1.25-in. eyepiece. The true field of view in arcseconds is given as well. With all this data provided, it should be easy enough to compare what the simulations show with what you're seeing at the eyepiece.

Why NGC and SAO Numbers?

There are many different ways to refer to objects in the sky. The star Vega, for example, is referred to as Alpha Lyrae in Johann Bayer's star catalog, 3 Lyrae in John Flamsteed's, and HIP 91262 in the Hipparcos Catalog. Deep sky objects often have more than one common name as well as various catalog designations. The famously bright nebula in the sword of Orion has been called the Great Nebula and the Great Orion Nebula, as well as M42 and NGC 1976.

Go-to telescopes take this sort of thing in their stride, offering the user more than one way to nominate a particular target. In the case of the Orion Nebula, for example, you could find it by its Messier number, its NGC number, or by its common name. Although all these methods work equally well, bouncing around from one catalog to another requires a lot of button pressing. This can be very tedious, especially if it's cold and you want to put your gloves back on!

So, to avoid this problem, this book uses just two catalogs, both of which are available on all the major go-to telescope systems, including the popular Celestron and Meade models. These are the New General Catalogue (NGC), a list of 7,840 deep sky objects, and the Smithsonian Astrophysical Observatory Star Catalog (SAO), a list of 258,997 stars.

There's nothing to stop you using other catalog designations if you want to; it's just that by staying with these two catalogs you'll be able to turn your telescope to more objects more quickly, and with a minimum number of keystrokes.

Figure 1.1. Most go-to telescopes, including those in the Celestron NexStar range, can find deep sky objects and stars by their NGC and SAO numbers. Sticking with just these two catalogs opens up the widest range of targets with the least amount of button pushing. (Photo courtesy of Celestron.)

Observing from the Suburbs and Exurbs

There are two main problems with observing from suburbs: light pollution and the built-up horizon.

So far as backyard astronomy goes, there isn't much you can do about trees and buildings along your horizon, short of moving your house. For northern hemisphere observers, it's the southern horizon that matters most, since that's where the constellations close to the celestial equator are to be found. The situation is reversed for southern hemisphere observers.

In either case, if you can't see that horizon clearly, there will be an important part of the sky you cannot see no matter what the season. The other horizons don't matter nearly so much. Why? Because even if a constellation is close to that particular horizon for one season, it'll be higher up in the sky elsewhere in the year.

While annoying, a built-up horizon isn't a disaster; the main thing is you plan your observing session to take into account whatever stars and deep sky targets are favorably placed at that particular time of year.

Light pollution is a much more serious and difficult-to-fix problem. Light pollution is the upwelling light from cities that drowns out faint stars and deep sky objects. To get the best views of the night sky, you really do need to visit a dark sky site. This usually means driving 50 or a 100 km away from home, and while that might be fun once in a while, it isn't really practical in the long term. So for most amateur astronomers living in the suburbs, light pollution is something that has to be understood and dealt with.

The standard scale for measuring light pollution is called the Bortle Dark Sky Scale. It runs from 1 to 9, with 1 being a perfectly dark sky and 9 being a grossly light-polluted inner city sky. Most suburban skies will rate somewhere between 6 (a bright suburban sky) and 4 (dark suburban skies transitioning to rural skies).

If your skies are at Class 6, the limiting magnitude of stars visible to the naked eye will be between 5.1 and 5.5. The Milky Way will be visible only directly overhead, and even then, it'll be very washed out. The horizon up to 35° will be suffused with glow from nearby urban areas, and few deep sky objects will be visible to the naked eye, likely only the Pleiades and the Orion Nebula, and possibly also the Andromeda Galaxy and the double cluster in Perseus.

A Class 5 sky is, of course, darker, and the Andromeda Galaxy should be readily apparent to the naked eye, but the nearby galaxy M33 isn't visible. The limiting magnitude for stars is 5.6–6.0. The Milky Way should be relatively clear and easy to see not just overhead but also a good two-thirds of the way down to the horizon.

At Class 4, the Milky Way can be seen over a large arc of the sky, almost to the horizon, but it lacks the visual oomph it has at a rural or dark sky site. The naked eye limiting magnitude is between 6.1 and 6.5. The benchmark deep sky object here is the galaxy M33 in Triangulum; this should be just about visible, with averted vision, in a Class 4 sky.

Not all suburban astronomers live in suburbs; many live in what geographers call exurbs. These are perhaps better known as dormitory or commuter belt towns, often built in the country but connected to a nearby city by highways and fast railway links. These can be excellent places to do your backyard astronomy from. Although light pollution will still be present, it is usually much less of an issue than in the suburbs proper. If nothing else, the worst of the light pollution tends to be concentrated along the horizon in the direction of the nearest big city.

A typical exurb sky will rate a 4 or 3 on the Bortle Dark Sky Scale. At Class 3, such a sky will enjoy a limiting magnitude of between 6.6 and 7.0. Light pollution will be most noticeable close to the horizon, but overhead the sky will be relatively dark. The Milky Way will be sufficiently bright that its shape can be seen and several deep sky objects should be visible to the naked eye, including M33 as well as the brighter globular clusters such as M13, M15, and M22.

All the objects in this book should be visible from sites with skies between Class 6 and Class 3. For the most part, objects are described as they appear from the author's main observing location in Hertfordshire, and as such, refer to skies somewhere around Class 5.

Light Pollution Filters

Light pollution filters are moderately priced accessories well worth adding to your toolkit. They work by blocking certain wavelengths of light, thereby making deep sky objects easier to see. There are three basic types of light pollution filter, and choosing which is best for you will depend on the types of objects you want to observe.

Depending on its type and size, light pollution filters range in price from around $75 to over $200. They can be bought in one of two sizes that screw into the threads inside the barrels of almost all 1.25-in. and 2-in. eyepieces, including all the big name brands. Before buying a light pollution filter, think carefully about which size you'll want. Planetary nebulae, for example, are usually viewed at medium to high magnification levels, so a 1.25-in. eyepiece is likely what you'll be using to view such objects. But large emission nebulae are often best viewed at low powers and through wide field eyepieces, in which case a 2-in. eyepiece would be best.

Light pollution filters are also available as units designed to screw onto the rear cell (the opening) at the back of the optical tube of SCTs and Maksutov–Cassegrain telescopes. These are mostly used by astrophotographers who need to attach cameras directly to the optical tube, but visual astronomers can use them, too. When in use, these rear cell filters connect to whatever star diagonal you want, and as such are compatible with both 1.25-in. and 2-in. eyepieces. On the other hand, it quickly becomes a chore screwing and unscrewing the star diagonal and the rear cell filter each time you want to swap between viewing with a filter and viewing without a filter.

There are three types of light pollution filters: broadband filters, narrowband filters, and line filters.

Broadband Filters

Broadband filters, sometimes called wideband filters, are designed to cut out skyglow but very little else. Some skyglow occurs naturally, and even in very dark sky sites some astronomers find broadband filters useful for observing very faint objects. But manufacturers usually sell broadband filters on account of their ability to cut out artificial sources of skyglow, i.e., light pollution. That alone should make them very popular among suburban astronomers.

But do they work? Yes. Broadband filters help block the light emitted by street lighting, including high-pressure sodium lights and mercury vapor lights. If these types of lighting are particularly prevalent and annoying in your area, then a broadband filter might be worthwhile. Furthermore, unlike narrowband and line filters, broadband filters do not block starlight. This makes them usable when observing objects composed of stars, i.e., galaxies, globular clusters, and open clusters. They

Figure 1.2. Broadband filters, such as the Celestron LPR, are useful primarily in situations where artificial light pollution is slight. (Photo courtesy of Celestron.)

can also be used for observing most reflection nebulae, since these don't emit light in themselves but only reflect starlight from a nearby star.

However, in almost all situations, narrowband or line filters provide much better views of planetary and emission nebulae, as well as most supernova remnants. Even when used for viewing objects that should show some benefit, such as galaxies and globular clusters, broadband filters only improve things very slightly. Experienced amateur astronomers often find that increasing the magnification provides about the same degree of improvement in terms of dimming the background sky, rendering a broadband filter more or less redundant. Such astronomers usually find themselves using narrowband filters much more often, even though on paper at least such filters are useful on a smaller variety of objects.

In general, then, when it comes to broadband filters the best advice is to try one out before you buy one. Most astronomy clubs will have one to hand, so stopping by your local club and testing a broadband filter out on a variety of deep sky objects will give you a better sense of how useful such a filter might prove to be.

Narrowband Filters

Compared with broadband filters, narrowband filters block more wavelengths of light. As well as blocking street lighting and skyglow, they also block incandescent and fluorescent lighting as well, making them especially effective tools for use under suburban skies. However, unlike broadband filters, narrowband filters block starlight, so they are not used when observing objects that emit or reflect starlight, i.e., reflection nebulae, open clusters, globular clusters, and galaxies.

What narrowband filters typically do is let through three particular wavelengths of light: 486, 496, and 501 nm. Virtually everything else is blocked. Why these three wavelengths? Doubly ionized oxygen, often called O-III, emits light at 496 and 501 nm, and ionized hydrogen of a type called hydrogen beta (or H-beta) emits light at 486 nm. Both O-III and H-beta are characteristic of emission and planetary nebulae, and by letting through only the light from these ionized gases, those particular deep sky objects become much easier to see.

One commonly voiced criticism of narrowband filters is that they dim the view to such a degree than they cannot be used with small aperture telescopes. This is simply not the case, and narrowband filters can be extremely useful even with very small telescopes, provided they are being used to view an appropriate target. Yes, narrowband filters dim stars, but you shouldn't be using a narrowband filter if you want to observe stars or deep sky objects that are composed of stars.

In fact, narrowband filters are probably the most rewarding filters in terms of cost compared to usefulness. Although a bit more expensive than broadband filters, they are much more likely to get used, and the resulting images of emission and planetary nebulae will be significantly better. Most experienced amateur astronomers agree that if you only have the funds for one light pollution filter, then a narrowband filter will be the one to choose.

Line Filters

Line filters let through even less light than narrowband filters. The two filters of this type that are most widely used by amateur astronomers are O-III and H-beta filters.

Figure 1.3. Line filters such as this Celestron O-III provide useful benefits to suburban astronomers observing certain planetary nebulae, emission nebulae, and supernova remnants. (Photo courtesy of Celestron.)

As you'd expect, the O-III filter lets through the two wavelengths of light at 496 and 501 nm associated with doubly ionized oxygen, while the H-beta filter only lets through light at a wavelength of 486 nm.

Compared with narrowband filters, line filters can improve the views of certain deep sky objects significantly. In some cases, the improvement is so great that an otherwise invisible object becomes visible. The classic examples are the Veil Nebula in Cygnus (where an O-III filter handily outperforms a narrowband filter) and the Horsehead Nebula in Orion (which is essentially invisible without an H-beta filter).

On the whole, line filters aren't essential purchases in the same way that a narrowband filter probably is. But that said, there are objects for which one or other line filter is especially useful, as will be outlined in the entries for such deep sky objects in the chapters that follow.

Using Light Pollution Filters

Light pollution filters are easy to use, but one often neglected quirk is that they work best, i.e., provide the most contrast, when used at (or close to) a particular exit pupil value. The precise value will depend on the filter type, and manufacturers may supply this information with their products. But in general, an exit pupil between 3 and 7 mm is typically recommended.

To calculate the exit pupil, simply divide the f-number of your telescope with the focal length of the eyepiece. In the case of a standard-issue f/10 SCT, if you were using a 25 mm-eyepiece, then the exit pupil would be $10/25 = 2.5$. That's a small exit pupil and is clearly a bit on the low side, so you wouldn't get best results from a light pollution filter if used with this eyepiece. Swap the 25 mm eyepiece for a 32 mm eyepiece and things improve. The exit pupil rises to 3.2, within the optimal range.

One common use for narrowband and line filters is what might be called 'blinking.' This is where the filter is not screwed into the eyepiece barrel, but instead held between the eye and the eyepiece. By moving the filter in and out of your line of sight, you should be able to notice the stars alternate between being dimmed and not dimmed. What's the point of this? Imagine you know there's a small planetary nebula in the field of view, but you can't pick it out from the stars that surround it. With the filter in your line of sight, the stars will be dimmed, but the nebula won't be. As you shuffle the filter back and forth, the nebula will 'blink' into sight.

Reducer-Correctors

Reducer-correctors are screw-on lenses that shorten the focal length of most SCTs while flattening the field. As such, they correct two shortcomings that are particular to telescopes of the Schmidt–Cassegrain design: SCTs tend to produce narrow fields of view because of their long focal lengths, and because they produce slightly curved fields, stars towards the edge of the field appear less sharp than those at the center.

All you do is screw the reducer-corrector onto the rear cell of the telescope, and then attach whatever star diagonal and eyepiece you want to use. A standard 200 mm SCT will be converted from an f/10 telescope into an f/6.3 telescope, making it much more suitable for viewing objects with large angular diameters such as emission nebulae and open star clusters. Without a reducer-corrector, the widest field of view possible when a 1.25-in. eyepiece is used with a 200 mm SCT is 50 arcminutes, which is what you'd get with a 32 mm-Plössl. Add an f/6.3 reducer-corrector to the mix and that same eyepiece will produce a much more generous 1.3° field of view. Replace that 1.25-in. eyepiece with a 2-in. eyepiece, and the maximum field of is increased still further: a 35 mm-Panoptic, for example, will produce a whopping 1.9° field of view.

Figure 1.4. Reducer-correctors are useful for shortening the focal length of an SCT while also flattening the field. This makes such telescopes much better tools for observing deep sky objects such as open clusters that have a large angular diameter. (Photo courtesy of Celestron.)

Are reducer-correctors worth buying? Very definitely yes! For the suburban astronomer using a go-to SCT, these relatively inexpensive accessories are second only to light pollution filters in usefulness.

Both Celestron and Meade sell f/6.3 reducer-correctors suitable for both photographic and visual use; there are also f/3.3 reducer-correctors available for photographic use only. Although there may be slight cosmetic differences between the Celestron and Meade units, performance seems to be identical, and because they are designed to fit the standard rear cell thread at the back of 200 mm SCTs, you can use both reducer-correctors on both brands of telescope.

The Celestron and Meade f/6.3 reducer-correctors are fully compatible with all the popular 200 and 125 mm SCTs sold by the two companies. They can also be used with larger aperture SCTs, though vignetting can be a problem when long focal length 2-in. eyepieces are used. Celestron and Meade f/6.3 reducer-correctors may be used with Meade LX-series Maksutov–Cassegrain telescopes, but they are not compatible with the Meade ETX-series of Maksutov–Cassegrain telescopes.

Dark Adaptation

One important issue for suburban astronomers to consider is dark adaptation. It will take your eyes some time to become fully adapted to seeing in the dark. Once your eyes have become fully dark adapted, you will be able to see much fainter objects than otherwise. This means not only that difficult to see things such as galaxies will be clearer, but also that detail on objects such as globular clusters will be easier to discern.

The length of time it takes to become almost perfectly dark adapted is about 40 minutes, though most people will find their dark vision good enough to use profitably after just 20–30 minutes. You can spend that time aligning your telescope, and then move on to those objects for which dark adapted vision isn't important, such as planets, double stars, or the brighter open clusters. The Moon is so bright it will prevent dark adaptation, but if the Moon is up, you're probably not going to be tracking down faint deep sky objects anyway!

Some amateur astronomers maintain that it takes an hour or more to become fully dark adapted, but there's no experimental evidence that this is the case. What might happen, though, is that as the night wears on, you get better at discerning detail in faint objects. The sky likely gets a bit darker, too, as the last of the twilight vanishes. On the other hand, if you're getting cold and tired, you're going to see less, or at least you'll expend less effort looking for detail. Once that happens, it's time to go back indoors!

Oddly, perhaps, while it takes half an hour for your eyes to become accustomed to the dark, they re-adjust to bright light almost instantly. It's therefore important to make sure there aren't any local sources of bright light that might trigger this reaction. Switch off security lights, for example, and draw the curtains in any rooms with lights on. It should also go without saying that you shouldn't go indoors to look at a map or computer screen, either.

If you need a source of light while you're outdoors, stick with a low intensity red light. Various flashlights of this type are sold for use by amateur astronomers, typically using red LEDs or similar. Some planetarium programs can switch laptops over to a night vision mode that consists of dim shades of red, and when used like this they shouldn't compromise dark adaptation too much.

At least some go-to telescopes have small lamps built into their handsets; on the Meade Autostar handsets this is called the utility light and can be switched on and off using one of the buttons on the handset.

Getting the Most from a Go-To Telescope

Go-to telescopes work by using motors to adjust the position of the optical tube such that it points towards the desired object based on the coordinates calculated by the built-in computer. Obviously, the telescope can't see what it's looking at, and its accuracy will be only so good as the mechanics allow it to be. There are essentially four limiting factors here: the accuracy of the initial alignment; the stability of the tripod; the tightness of the clutches; and the health of the batteries. If you want to get the very best from your go-to telescope, you have to consider all four of these factors and adjust them as necessary.

Let's start with the initial alignment of the telescope. The details will vary from model to model, but all go-to telescopes fall into one of two basic types: those that are aligned manually, and those that do so automatically. Increasingly the trend is towards telescopes that align themselves automatically, using data from global positioning satellites (GPS) and their own built-in electronic compass. Once switched on, the telescope obtains time and location data from the satellite and uses its compass to find out which direction is north. From there, it slews itself towards two bright stars, at which point the user fine-tunes the alignment to make sure the star appears dead center in a medium to high magnification eyepiece. With that done, the telescope is good to go!

Older manually aligned go-to telescopes required the user to enter the time, date, and location data. In the case of the Meade systems, the user also had to put the telescope into the Home position with the tube horizontal and pointing north (in the northern hemisphere) or south (in the southern hemisphere). From there, the telescope would slew to two bright stars, the user would center them in the eyepiece field of view, and then the telescope would be aligned.

Celestron systems worked differently, with the user slewing the telescope to two or three bright stars, depending on whether two-star alignment mode or Sky-Align alignment mode was being used. The computer would calculate the stars most likely to fit that pattern, and from there finish off the alignment process. (At the time of this writing, some go-to telescope models in both the Celestron and Meade ranges retain manual alignment, but it's probably fair to assume that over time these will be replaced by GPS-equipped ones.)

So what can go wrong during the alignment process? The GPS-equipped go-to telescopes are essentially problem-free, and providing you follow the instructions there isn't much to go wrong. Weak batteries can cause problems, and as can loose or dirty connectors at the ends of the cable connecting the handset to the telescope.

Manually aligned telescopes (obviously!) won't work properly if you enter the wrong time and/or date, or forget to toggle the appropriate setting for daylight savings (summer) time. These errors are astonishingly easy to make, and if you find your go-to telescope consistently fails to locate objects, reviewing the time, date, and perhaps the location data are all troubleshooting steps well worth doing. In the case of the Celestron systems, it's also important you choose bright stars that are far apart from each other, and in the case of Sky-Align mode, that the three bright star 'objects' chosen are arranged in a triangular shape rather than a straight line. On the Meade systems, the computer will automatically choose appropriate alignment stars, so this shouldn't be an issue. Celestron telescopes also have Auto-Two Star and Two Star Alignment modes that work in a similar way, although not used automatically.

Make sure that the tripod is as level as possible. Adjust the length of the legs a little if you need to. Doing this by eye is normally adequate, but the Celestron telescopes come with built-in spirit levels (bubble levels) that will help.

Both manually aligned and GPS-equipped go-to telescopes will lose their alignment if the tripod is wobbly or otherwise destabilized. Kicking the tripod can knock the alignment out, and this is especially the case with telescopes mounted on relatively lightweight aluminum tripods. In fact all telescopes benefit from heavy, stable tripods, whether go-to or not, but because the computers inside go-to telescopes assume the telescope is on a static mount, it is doubly important that the tripod be as steady and stable as possible.

Go-to telescopes have clutches that connect the optical tube to the motorized gears that move it about. On some telescopes these clutches can be loosened so that the telescope tube can be moved manually, in which case the clutches will need to be tightened again before it will work in go-to mode. It is not easy to adjust the tightness correctly, and this sometimes causes the optical tubes to slip; even tiny amounts of slippages can mess up the ability of the telescope to find and track objects.

A common mistake is to tighten the clutches adequately for a lightweight eyepiece, but as soon you switch to a heavier eyepiece, the looseness of the clutch allows the optical tube to start slipping. So if your evening starts off fine, but over time the telescope gets less and less good at finding its targets, check the clutches. Sometimes tightening them while the telescope is pointing at a target will do the trick, but don't tighten the clutches while the telescope is slewing between targets. The instruction manual that came with your telescope will detail how to tighten these clutches and by how much.

The final source of problems is the set of flashlight batteries inside your telescope. Go-to telescopes are power hungry machines, and lack of power is probably the single most common reason why they stop working reliably. Old batteries may have enough power left to get the telescope aligned, but as the observing session proceeds, pointing and tracking becoming steadily less accurate. Cold conditions can have a similar effect, by slowing down the chemical reactions inside the battery that produce the power your telescope needs.

In short, although flashlight batteries are adequate for casual use, hobbyists who spend more than a couple of hours observing, or have to put up with very cold nighttime conditions, will find flashlight batteries unreliable and expensive. There are two alternatives here: external power packs and mains adapters.

External 12-V battery packs typically contain heavy duty lead-acid batteries that are recharged every 12 hours or so. These will power a go-to telescope for many hours and are very tolerant of cold conditions. Although a little more expensive than a set of flashlight batteries, these power packs will last much longer, and ultimately pay for themselves many times over.

Mains adapters are potentially useful and cheap to use, but using anything connected to mains voltage in a wet environment such as a garden is a risky business. Inexpensive mains adapters don't always produce the stable voltage that delicate electronics such as computerized telescopes need, leading to problems with

Figure 1.5. According to their manufacturers, go-to telescopes fail to work properly most often because of weak batteries. Consider replacing the flashlight batteries that go inside the telescope with a much more durable and reliable external power pack, such as this Celestron unit. (Photo courtesy of Celestron.)

reliability, and perhaps even causing damage. So it's important to choose a mains adapter specifically designed for use with telescopes in an outdoor environment.

Finally, the cables and connectors can cause all sorts of problems. Check that all cables are securely connected, and that all the ports and connectors are clean, dry, and dust-free. Damaged cables will prevent a go-to telescope from functioning properly, so these need to be looked after carefully between observing sessions. The author recalls sending a brand new go-to telescope back to the retailer at great expense, only to find out that the cable was faulty and needed replacing!

CHAPTER TWO

Winter

Although winter nights are very cold, amateur astronomers have good reason the bundle themselves up into warm clothing to brave the chill. For one thing, the ecliptic is at its highest this time of year, so if you're observing the planets, they'll be well above the murky skies of the horizon, resulting in the best possible views. But the wintertime sky also contains some of the most spectacular deep sky objects, of which the Orion Nebula and the Pleiades are the best known.

It shouldn't be overlooked that this is also the time of year when people give each other gifts. Quite a few budding astronomers will be testing out their first telescope under wintertime skies, and as we'll see, there are some real treats in store for them!

The winter sky is dominated by the Milky Way, and most of the objects observed at this time of year are within the Milky Way, such as nebulae and open star clusters. In particular, we're mostly looking at things in one particular part of the Milky Way Galaxy, known as the Local Spur or Orion Arm.

Our galaxy is a barred spiral galaxy with four major arms and several minor arms. The four major arms are known as the Perseus Arm, Scutum-Crux (or Centaurus) Arm, the Cygnus-Norma Arm, and the Sagittarius (or Sagittarius-Carina) Arm. As you'd expect, their names come from constellations within which their particular bits of the Milky Way passes through. Of the four major arms, the Perseus arm and the Scutum-Crux arm are by far the largest. If the Milky Way was viewed from above, these would be the two arms that would catch the eye first, the two other major arms slotting inside them.

As well as the four major arms, there are several fragmentary minor arms sometimes called spurs. One of these is the Orion Arm, and this is where our Solar System lies. Being relatively nearby, it contains many of the biggest and brightest deep sky objects, including the Orion Nebula, the Pleiades, the Dumbbell Nebula, the

Beehive Cluster, the Hyades, the Ring Nebula, and the Owl Nebula. On the other hand, because we're mostly looking into the Milky Way rather than away from it, objects outside our galaxy don't figure much this season, with a few notable exceptions such as the Andromeda Galaxy.

Suburban astronomers will find winter a particularly good time of the year for deep sky observing. The objects visible during this season are mostly those that are either unaffected by moderate levels of light pollution, as in the case of open star clusters, or else benefit from the use of light pollution filters, as is the case with emission nebulae and planetary nebulae.

Of course there is a downside to wintertime observing: the weather! Or more specifically, the freezing cold weather characteristic of this time of year. To get the most out of wintertime observing you'll need to be wearing suitably warm clothing. This will usually mean several layers on your body, a hat to keep your head warm, and a pair of gloves. In the colder parts of the world, this might be augmented with earmuffs and a scarf.

It's also important to take your telescope outside before you want to start observing. This gives the telescope time to cool down to the ambient air temperature. Otherwise the warm air inside the tube will move about, messing up the images you see at the eyepiece. This is particularly noticeable at high magnifications, where you'll find stars flickering rather than forming small, stable points.

The amount of time it takes for a telescope to cool down properly will depend upon how big the telescope is, and how much colder the air outdoors is compared with the temperature inside your home. Open tube telescopes such as Newtonian reflectors tend to cool down quickly because the warm air can escape quickly, whereas closed tube telescopes such as SCTs take much, much longer. Something like a 200-mm SCT will typically take 1–2 hours to completely cool down, whereas a reflector of similar aperture will be ready within half an hour.

Showpiece Objects

NGC 224 (M31, Andromeda Galaxy)

Of all the galaxies in the northern sky, none rival the Andromeda Galaxy in size or brightness. At a mere 2.5 million light years away, it is far closer to us than any other major galaxy, and its angular size is about six times that of the full Moon. But the Andromeda Galaxy is more apt to disappoint than any other object in the sky (with the possible exception of the planet Mars which, most of the time, is much smaller and reveals less detail than its almost mythic reputation would suggest).

The main reason the Andromeda Galaxy is a 'difficult' target is that it is so big. At about 3° in angular width, you need a very wide field of view if you want to see the whole thing in its entirety. In practice very few telescopes other than short focal length refractors or Newtonian reflectors provide anything like the necessary fields of view, and certainly not the popular SCT and Maksutov telescopes, around which most go-to telescope systems are built. So for the most part, amateur astronomers end up viewing just a bit of this galaxy at any one time.

With an apparent magnitude of about 4, the Andromeda Galaxy is technically quite a bright object, but its sheer size means that the light it emits is rather spread out. As a result, it doesn't stand out from the background sky as much as more compact objects of similar brightness might do. In fact at first glance, you'll probably only notice the central core. It isn't difficult to see the Andromeda Galaxy then, but between its size and its lack of contrast, it is a challenge to get a good view of the whole thing.

So what can you actually see? A 200-mm SCT fitted with a 32-mm Plössl delivers a field of view of about 50 arcminutes (or put another way, 5/6 of a degree). This is enough to show the core of the Andromeda Galaxy, which should be visible as a fairly bright, rather large elliptical blur. It looks like a cloud, and it isn't difficult to see why such objects were called nebulae, from the Latin word for 'cloud.' In fact, older astronomy books commonly referred to this galaxy as the Great Nebula in Andromeda or words to that effect, and continued to do so at least into the 1950s, long after it was demonstrated that it wasn't a nebula at all but a very distant collection of stars much like our own Milky Way.

One difference between galaxies and nebulae is the effect light pollution filters have on them. Narrowband and line filters aren't useful at all, and broadband filters help only up to a point. Broadband filters will block natural skyglow and a small amount of artificial light pollution, but won't produce anything like the view you'd get of this galaxy when observed from a genuinely dark site.

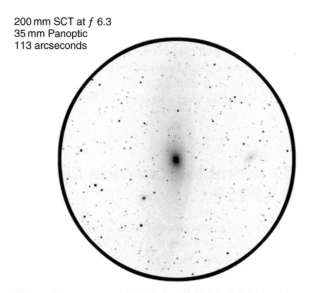

200 mm SCT at ƒ 6.3
35 mm Panoptic
113 arcseconds

Figure 2.1. The Andromeda Galaxy is notoriously disappointing despite its size and relative brightness, but you should at least detect its bright core, the smudgy spiral arms around the core, and given reasonably dark conditions, its two satellite galaxies. Ample aperture and a wide field of view are both helpful. (Image produced using Starry Night Pro. AllSky data courtesy of Main-Sequence Software Inc.)

Is the Andromeda Galaxy a total waste of time then? Far from it! For a start, have a look around for its two companion galaxies, M32 (NGC 221) and M110 (NGC 205).

Assuming you're using a low-powered eyepiece centered on the core of the Andromeda Galaxy, then M32 should be easily visible as a near-spherical blob of light towards the edge of the field of view, 22 arcminutes away from the center of the Andromeda Galaxy. M32 is surprisingly massive for its size and type and probably has a supermassive black hole at its center. It is also rather odd in lacking globular clusters; most galaxies have at least a few; its neighbor, M101, has at least eight, and the Andromeda Galaxy around 460.

M110 is a bit farther out, about 36 arcminutes from the core of the Andromeda Galaxy. It is a reasonably bright elliptical blob, and while still classified as a dwarf elliptical galaxy, it's about twice the size of M32. Both M110 and M32 can be mistaken for each other.

What else is there to see in the Andromeda Galaxy? Really big telescopes can spot star clouds within this galaxy, most notably NGC 206, and under good, dark skies dust lanes can be seen as well. But under suburban skies none of these features are going to be visible.

Instead, the Andromeda Galaxy is an object best appreciated in the imagination. The light that arrives at your eyepiece left the Andromeda Galaxy 2.5 million years ago, about the time the earliest human-like animals, *Homo habilis*, were walking about and using stone tools. The Isthmus of Panama had only recently formed, and the great faunal interchange of mammals between the North and South American continents was starting to take place. In geological terms, these events took place almost yesterday – the dinosaurs, for example, having lived between 230 and 65 million years ago. But that's the point: as distant as 2.5 million years seems, and as far away as 2.5 million light years might be, these are actually very small numbers when measured against a geological or cosmic scale.

The Andromeda Galaxy may seem a long way off, but it's actually in our galactic backyard.

NGC 1502 (Kemble's Cascade Cluster) – See Also SAO 12969

Kemble's Cascade is a very pretty chain-shaped asterism of about twenty reasonably bright stars in the otherwise unexciting constellation of Camelopardalis. At one end of the chain is the open star cluster NGC 1502, sometimes called the Kemble's Cascade Cluster.

NGC 1502 is an open star cluster of around 45 moderately bright stars. At its center is a particularly attractive double star known as Struve 485, made up of two seventh-magnitude stars, SAO 13030 and SAO 13031. These two stars are separated by about 18 arcminutes. NGC 1502 is a relatively young cluster, about 11 million years old, and located about 3,000 light years from Earth.

NGC 1535 (Cleopatra's Eye)

This planetary nebula is one of very few easy targets in the constellation of Eridanus, a southerly constellation often overlooked by astronomers at mid to high northern latitudes. From southern England, for example, this object doesn't rise much more than 25° above the southern horizon. But it's well worth trying to pin down. For a start, like most planetary nebulae, it's small but bright, and benefits greatly from the use of light pollution filters. Both narrowband and O-III filters work well on this object. Boosting magnification heightens the contrast further, and while you might not see much detail, you will at least be able to tell that it's a planetary nebula rather than a star.

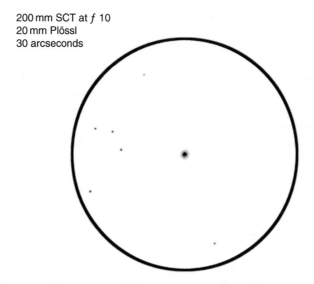

```
200 mm SCT at ƒ 10
20 mm Plössl
30 arcseconds
```

Figure 2.2. Use a narrowband or O-III filter to distinguish Cleopatra's Eye (at center) from the very similar looking stars in the same field of view. (Image produced using Starry Night Pro. AllSky data courtesy of Main-Sequence Software Inc.)

Under good conditions this nebula is one of that select group of deep sky objects that has some color, in this case a vague blue-green hue. You will need a reasonably large telescope for this to be obvious, though, around 200 mm at minimum. Large aperture telescopes, upwards of 250 mm, are required to make out the magnitude-12 star at the heart of the nebula. NGC 1535 is about 1,500 light years from us.

NGC 1976 and 1982 (Orion Nebula)

If you need one good reason to brave the winter chill on a cold December night, then this is it. Few deep sky objects rival the Orion Nebula in beauty, brightness, and interest, and none surpass it. It really is the standard against which all other deep sky objects are compared!

When you turn your telescope towards the Orion Nebula, you're looking deep into something called the Orion Molecular Cloud Complex. This vast cloud of dust and gas is some 240 light years in width, and includes not just the Orion Nebula but also the Flame Nebula (NGC 2024) and the notoriously elusive Horsehead Nebula (IC 434). The Orion Molecular Cloud Complex is a relatively nearby object, lying in the same minor arm, the Orion Arm, as our own Solar System. This is why the thing is so big and, in some parts at least, so incredibly bright.

The Orion Nebula is a region of the Orion Molecular Cloud Complex that is especially nearby, a mere 1,300 light years away. It is visible even to the naked eye, as a fuzzy patch halfway down the Sword of Orion asterism. A small telescope reveals its two brightest regions, the larger of which is M42 (NGC 1976) and the other M43 (NGC 1982).

The Orion Molecular Cloud Complex is generally a site of active star formation, and the Orion Nebula is no exception. Long-exposure photographs reveal all sorts of complex features, including star clusters, reflection nebulae, and clouds of gas and dust. The Hubble Telescope has also shown evidence of protoplanetary discs, known as proplyds, meaning that at least some of the young stars in the Orion Nebula are well on the way to forming their own solar systems.

Amateurs may not see that level of detail, but they can still enjoy the view. The Trapezium Cluster, for example, is a good starting point. This celebrated cluster of four bright stars is a very young cluster, a mere million years in age. Although only four stars are immediately visible, and one or two more with more magnification and good seeing conditions, there are actually many more stars in the cluster obscured by the surrounding dust and gas. Infrared images have allowed astronomers to see these stars, and in fact the Trapezium Cluster has become one of the best studied of all open star clusters.

Although a broadband filter will improve your views of the Orion Nebula some-what, a narrowband filter will be much more useful. The nebula typically looks bigger when viewed with a filter because the outlying parts become easier to see. Of course a narrowband filter will also dim the stars within the nebula, and one reason a broadband filter might be preferred is so that you can view both the nebula and the stars at the same time. O-III filters are useful, too, and there are subtle differences in how the Orion Nebula appears when viewed with O-III filters and narrowband filters. An H-beta filter isn't the filter of choice for viewing the overall Orion Nebula, but it does seem to bring out the details on the M43 part of the nebula in a way the other filters don't.

Unlike most deep sky objects, long-exposure photographs don't provide better views of this object than you get through a telescope. Because the Orion Nebula has regions that span a tremendous range of brightnesses, long-exposure photographs tend to show the fainter extensions rather better than you see at the eyepiece, but overexpose the brighter core of the object. By contrast, your eyes may not be as sensitive to dim objects as cameras, but they're much better at dealing with varying levels of brightness. So while you won't see quite as much detail around the edges of

200 mm SCT at ƒ 6.3
35 mm Panoptic
113 arcseconds

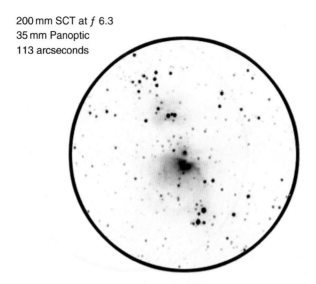

Figure 2.3. The Orion Nebula is big, bright, and easy to explore with all types of telescopes and across a range of magnifications. (Image produced using Starry Night Pro. AllSky data courtesy of Main-Sequence Software Inc.)

the Orion Nebula, the view you'll see at the eyepiece is at least as good as any photo. Indeed, some seasoned observers think it's rather better.

The glowing gas that you see is what is known as an emission nebula, a cloud of gas that fluoresces thanks to the young, hot stars within it. Science fiction films often give the impression that nebulae of this type look something like the water vapor clouds in Earth's atmosphere. However, these clouds of gas are incredibly tenuous, and by comparison with the air around us, are practically vacuums! But because they're so big, and the stars inside them are so hot, emission nebula can be very impressive objects when viewed from our distant vantage point.

By far the bigger of the two halves of the Orion Nebula is M42. This bird-shaped cloud is something like 25 light years in diameter at its widest point, or about nine times the distance from the Sun to Alpha Centauri. M43 is the much smaller cloud of gas. Although they look like separate objects, they're actually just the one big cloud of gas, with seeming division between them being a vast lane of dust that obscures the light emitted by that part of the nebula behind it.

NGC 2168 (M35) and NGC 2158

M35 is a fantastic open star cluster in the zodiacal constellation of Gemini, and surely among the finest open star clusters in the northern sky. It's an easy object for any telescope, and the longer you look at it, the more stars you'll find within the cluster.

200 mm SCT at ƒ 6.3
35 mm Panoptic
113 arcseconds

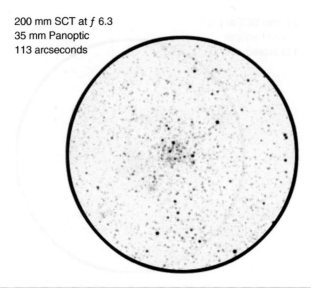

Figure 2.4. The bright open cluster M35 is easy to observe, but you need to look carefully to spot the smaller and fainter open cluster NGC 2158, in this image below and slightly to the left. (Image produced using Starry Night Pro. AllSky data courtesy of Main-Sequence Software Inc.)

Almost all are hot blue stars, and only a single red giant is a confirmed member of the cluster.

Some would argue that a medium-sized telescope, around 75-150 mm, is the optimal size for viewing M35, since bigger telescopes show too many of the fainter stars too easily, resulting in something a bit less engaging. On the other hand, it's an object that tolerates moderate levels of light pollution rather well, and if a few of the fainter stars are hidden from view, what's left over is still very attractive.

M35 is a large object, almost 30 arcminutes in diameter, so you need to use a low-power eyepiece to see it at its best. On a 200-mm SCT, a 32-mm Plössl works well, getting the whole cluster comfortably into the field of view.

Just to one side of M35 is a smaller and fainter cluster, NGC 2158. Small telescopes may reveal this object as a faint, fuzzy patch, but larger telescopes, from about 200 mm upwards, will be able to resolve out some of its constituent stars, including numerous yellow and orange stars that are older and cooler than the blue ones that dominate M35. Although an attractive cluster in its own right, NGC 2158 is easily overlooked given its proximity to M35.

Actually, although the two clusters seem close together from our point of view, they're actually a long way apart. M35 is about 2,800 light years away, compared to a distance of about 16,000 light year for NGC 2158. The two clusters are also very different in terms of age. M35 is about 100 million years old, making it relatively young by open star cluster standards. NGC 2158 is much older, current estimates suggesting it is between 1 and 2 billion years in age. That's why it contains so many yellow and orange stars, which tend to be older than the hotter and brighter blue ones.

NGC 2287 (M41)

M41 is an open cluster in Canis Major not far from Sirius, the brightest star in the night sky. Because of its relatively southerly location, M41 is sometimes overlooked by far northern observers, but it's well worth taking the time to examine. Under the right conditions it is extremely attractive, as well as rather an interesting open star cluster.

All telescopes will reveal good views of this object, though as is often the case with the brighter open star clusters, moderate apertures seem to provide the best balance between showing lots of stars without showing too many all at once. But it is a big cluster, some 38 arcminutes in diameter, and a low-power, wide-field eyepiece is what you need more than anything else. If you have a 200-mm SCT, you'll find that an f/6.3 reducer-corrector really earns its keep on this object, allowing a 32-mm Plössl to center M41 against a good-sized chunk of sky. A wide-field 2-in. eyepiece such as a 35-mm Panoptic could be used instead, if you have a 2-in. star diagonal, to get a similarly expansive field of view.

M41 is around 200 million years old, and although it contains mostly blue stars, it also contains a fair number of red giants. This makes it an interesting contrast to M35 in Gemini, an open star cluster that is only half the age of M41 and contains just a single red giant. The two star clusters are otherwise fairly similar, both being about 25 light years in diameter, though M41 is a bit closer to Earth at a distance of 2,300 light years compared to 2,800 light years for M35.

NGC 2392 (Eskimo or Clown Face Nebula)

The Eskimo Nebula is one of the nicest planetary nebulae in the sky, and certainly among the best of the ones visible during the winter months. It is a small, bright object, and unusually among deep sky objects, it has some color to it. Some observers see it as green, others as blue. The bigger the telescope, the more obvious its color will seem, while very small telescopes (anything less than 100 mm) may show no color at all.

At low magnifications the Eskimo Nebula appears to be a somewhat fuzzy star, and it's easy to imagine why such objects became known as planetary nebulae: they can indeed look a bit like planets. With increasing aperture and magnification its details become apparent. At high levels of magnification the two-shell structure that gives this nebula its popular name can be seen. The inner shell is the 'face,' and the outer shell is the fluffy fur of the Eskimo's hood. In the case of a 200-mm SCT, try using a 10-mm Plössl for a magnification of ×200. If the two-shell structure isn't apparent when looked at directly, use averted vision to see if the details snap into view.

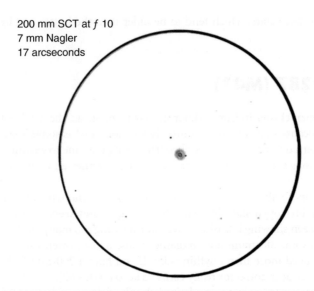

200 mm SCT at ƒ 10
7 mm Nagler
17 arcseconds

Figure 2.5. High powers bring out the best from the Eskimo Nebula, making its features easier to see and enhancing the contrast between the nebula and the background sky. (Image produced using Starry Night Pro. AllSky data courtesy of Main-Sequence Software Inc.)

With apertures upwards of 200 mm, you should be able to see a circular halo around a faint (ninth-magnitude) star as well. Long-exposure photographs reveal considerably more detail than this, but to see such details visually you need a very large telescope.

From our vantage point, we're actually looking down into a funnel of hot gas emanating from a dying star. The star itself was apparently rather similar to our own Sun, so when we look at the Eskimo Nebula, we're getting a glimpse of the likely fate of the Sun, a somewhat eerie thought.

Like other planetary nebulae, your views of the Eskimo Nebula benefit greatly from the use of a suitable filter. Light pollution filters help a bit, but your best views will come from using either a narrowband or O-III filter.

There is some debate over the distance between Earth and the Eskimo Nebula, values ranging from 1,600 to 7,500 light years, but most astronomers tend towards a value of about 3,000 light years.

The Eskimo Nebula gets its name from its appearance in photographs, which some have likened to that of a human face tucked up inside the hood of a parka jacket. Others see a clown's face, hence the name Clown Face Nebula.

NGC 2422 (M47), NGC 2437 (M46), and NGC 2438

These are a pair of superb open star clusters in the constellation of Puppis that mid to far northern observers tend to overlook because they don't rise far above the southern horizon. But they're well worth adding to your observing program, because they form such a nice pair of star clusters when viewed through a low-power, wide-field telescope. Of course they're nice objects when viewed on their own, but taken together, they make a particularly interesting sight.

Of the two, M47 is the easiest to see and under dark skies can be seen with the naked eye. It is fairly tolerant of light pollution and a good 20 to 30 stars should be seen through any telescope. M46 is much less easily resolved and can be difficult to spot under light-polluted skies. Through small telescopes it looks more like a hazy patch of light than an open cluster. When presented in the same field of view the contrast between the two clusters is immediately apparent. Although their size and proximity suggests that famous pairing of open clusters in Perseus we call the Double Cluster, in actual fact the view is very different indeed.

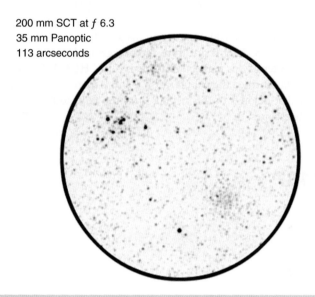

200 mm SCT at ƒ 6.3
35 mm Panoptic
113 arcseconds

Figure 2.6. M47 (top left) is easier to resolve than M46 (bottom right). Getting both clusters into the same field of view requires a very wide field, obtained here using a 2-in. eyepiece and a reducer-corrector. (Image produced using Starry Night Pro. AllSky data courtesy of Main-Sequence Software Inc.)

Rich-field refractors or fast Newtonians are the telescopes of choice for viewing the two clusters together, since the required field of view is a good 2°. By default that won't be possible using a standard 200-mm SCT with 1.25-in. eyepieces. But if you have an f/6.3 reducer-corrector and a wide-field 2-in. eyepiece such as a 35-mm Panoptic, the resulting view, assuming fairly dark skies and a clear horizon, can be spectacular.

To get the two star clusters into the same field of view you could have your go-to telescope slew to either star cluster, locate the other star cluster through the finder telescope, and then use the control pad to bring that second star cluster into the field of view. But an easier approach is to have the telescope center itself on the star SAO

153190 instead. This will place both M46 and M47 in the field of view, together with a smaller and much dimmer open star cluster called NGC 2423. This little cluster is at the third point of a right-angled triangle, with M47 at the right angle and M46 the farther away of the two points that form the rest of the triangle.

M46 is the far bigger of the two clusters, about 30 light years in diameter versus about 17 light years for M47. But it is also much farther away, and that's why it is so much more difficult to see and resolve. M46 is believed to be about 5,500 light years away, compared to a mere 1,600 light years for M47. Their ages are pretty similar, though, with M46 being about 100 million years old, and M47 a little younger, at about 78 million years of age.

Seemingly within M46 is a small planetary nebula, NGC 2438. It's actually just an object that happens to be between us and M46, and not a member of the M46 star cluster at all. NGC 2438 is not easy to see at all and may be completely invisible to those under light-polluted suburban skies or using small- to medium-aperture telescopes. You need at least 200 mm to spot this object, and the use of a narrowband or O-III filter makes picking it out from the background stars dramatically easier. The star in the center of this planetary nebula is called HD62166. Although it looks like it's the star that formed the nebula, it isn't, and is merely in the same line of site; the actual star that belongs to this nebula is far too faint to be seen with amateur astronomers' telescopes.

NGC 2437

See NGC 2422.

NGC 2451

NGC 2451 is a star cluster in Puppis scattered across a 1.7° of space and centered on the orange supergiant star c Puppis (SAO 198398). It contains about 40 stars, of which c Puppis, at magnitude 3.6, is by far the brightest. Although an easy open cluster to find with binoculars and very attractive through wide-field telescopes, its relatively southern location in the sky makes it tricky to spot from mid-northern latitudes, and far northern observers may not be able to see it at all.

NGC 2451 is about 850 light years from Earth and is believed to be about 40 million years old, which explains why it is filled with mostly hot blue stars. There is some debate about the nature of this star cluster, some astronomers maintaining that it is merely as asterism and not a true cluster at all. In other words, although there seem to be a lot of rather bright stars neatly arranged in a clump, they don't have a common origin and weren't born from the same cloud of gas and dust (as would be the case for a true open cluster). Instead, NGC 2451 is something more like Kemble's Cascade, just a chance alignment of stars that catch the eye.

NGC 2632 (M44, Praesepe, Beehive Cluster)

The only impressive deep sky object in the dim constellation of Cancer, M44 is an easy naked-eye object under dark skies that reveals itself as a small but distinct cloudy patch of light. It was one of seven 'nebulae,' the Latin word for clouds, described by the Greek astronomer Ptolemy in the book we now know best through its Arabic translation, the *Almagest.*

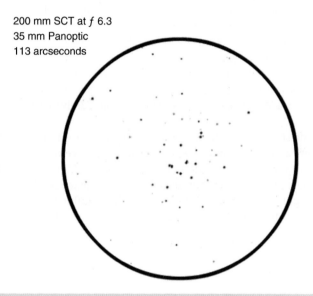

200 mm SCT at ƒ 6.3
35 mm Panoptic
113 arcseconds

Figure 2.7. The Beehive Cluster is best observed at very low powers, and the use of a reducer-corrector and a 2-in. wide field eyepiece is highly recommended. (Image produced using Starry Night Pro. AllSky data courtesy of Main-Sequence Software Inc.)

Ancient astronomers described this nebula as the manger from which two donkeys, the stars Asellus Borealis and Asellus Australis, were feeding. The word Praesepe is in fact the Latin for manger, and the names of the two nearby stars literally mean 'northern ass' and 'southern ass,' respectively.

The Greek poet Aratus, writing in the third century B. C., include the manger and the two donkeys in a famous poem, the *Phaenomena*, that describes various astronomical curiosities. One odd aspect of the poem is that he relates the appearance of the manger and the two donkeys to weather forecasting. For example, should the manger become dim but the two donkeys remain equally bright, then rain is likely. But if the manger and the northern donkey both become dim, and only the southern donkey stays bright, then you can expect winds from the south. This is a good reminder of how astronomy and astrology were connected in the past, the whole subject being one of practical rather than theoretical importance.

By 1844 the English astronomer (and naval officer) William Henry Smyth translated the name Praesepe as 'bee-hive' in his book on practical astronomy, *A Cycle of Celestial Objects*. Presumably the name referred to its swarm-like appearance; as Smyth pointed out, ancient astronomers were limited to naked-eye observations and couldn't see this object as anything other than a nebula. Smyth further explains that it wasn't until Galileo turned his telescope towards the Praesepe that it was seen to be made up of 36 small stars. Most astronomers now call this cluster the Beehive Cluster rather than the Praesepe.

If you want a good view of this object, which has an angular diameter of over 1.5°, you absolutely must use the widest-field, lowest-power eyepieces at your disposal. In the case of a standard f/10 200-mm SCT, a 1.25-in. eyepiece won't do justice to this object, though the view you'll get using a 32-mm Plössl will be pleasant enough. Instead, use an f/6.3 reducer-corrector to shorten the focal length, and a 2-in. visual back that allows for the biggest possible field of view. Suitably equipped, a 200-mm SCT at f/6.3 with a 35-mm Panoptic will return an incredibly lovely view of this open star cluster.

This star cluster is one of the closest ones to Earth. It was assumed to be about 525 light years away until quite recently, when its distance was revised upwards to about 575 light years thanks to additional data from the Hipparcos satellite. Only a few star clusters are any closer than this, the most notable ones being the Pleiades, around 440 light years away, and the Hyades, about 150 light years away (and do also see the entry for SAO 27876, a member of the very nearby Ursa Major Moving Group).

NGC 3201

This is one of the very few globular clusters visible in winter for northern observers, reaching its highest point above the southern horizon around midnight in mid-February. It is a fairly loose cluster by the standards of globular clusters, at Class X on the Shapley–Sawyer scale, and under good skies 100- to 150-mm telescopes will start to resolve this cluster nicely.

Unfortunately, while NGC 3201 is a lovely object by any standards, it is so far south that most northern hemisphere observers won't be able to see it. Even at latitudes as low as central Florida, NGC 3201 barely gets to more than 15° above the horizon.

NGC 3201 is about 16,000 light years from Earth and measures roughly 14 light years in diameter. Whereas most globular clusters orbiting the Milky Way Galaxy do so in the same direction as the rotation of the galaxy, NGC 3201 orbits the Milky Way in the opposite direction. Some astronomers have taken this to mean that NGC 3201 is a globular cluster that wasn't originally part of the Milky Way, but was instead captured by its gravity at some point in time. Although an intriguing possibility, this assertion is far from universally accepted.

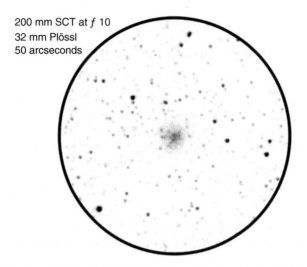

200 mm SCT at *f* 10
32 mm Plössl
50 arcseconds

Figure 2.8. NGC 3201 is a nice but rather southerly globular cluster, and one of the few good globular clusters visible at this time of year. (Image produced using Starry Night Pro. AllSky data courtesy of Main-Sequence Software Inc.)

Interesting Deep Sky Objects

NGC 205 (M110)

See NGC 224, the Andromeda Galaxy, in the showpiece objects section above.

NGC 221 (M32)

See NGC 224, the Andromeda Galaxy, in the showpiece objects section above.

NGC 752

This is a bright open cluster in Andromeda containing more than 60 stars, mostly white and blue in color, but some orange and yellow as well.

It's a veritable jewel box of a cluster, but because it spans some 50 arcminutes of space, you'll need the widest possible field of view to see the whole thing. On a 200-mm SCT, a 32-mm Plössl will just about squeeze the cluster into the field of view, but you'll get a better impression by using either a 2-in. eyepiece like a 35-mm Panoptic, an f/6.3 reducer-corrector, or ideally both.

NGC 752 is about 1,300 light years from Earth. It appears to be at least 1.1 billion years old, making it a rather old open cluster. (Most only last a few hundred million years until their constituent stars drift apart completely.)

NGC 1291

This elliptical galaxy in the southern sky constellation of Eridanus is not visible to observers at mid to high northern latitudes. Even as far south as central Florida it doesn't get more than about 16° or so above the southern horizon. That's a shame, because this galaxy is bright and easy to see even with small telescopes. Large aperture instruments should clearly show a bright core surrounded by a more hazy area around the outside. Very large telescopes (from 300 mm upwards) show hints of a huge ring of dust encircling the galaxy, a feature prominently shown in long-exposure photographs. In this regard, it is similar to the Sombrero Galaxy (NGC 4594).

NGC 1291 is about 30 million light years away and a member of the Dorado Cloud of galaxies.

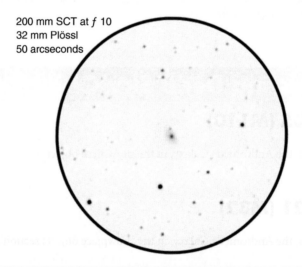

200 mm SCT at *f* 10
32 mm Plössl
50 arcseconds

Figure 2.9. NGC 1291 is a southerly galaxy that is invisible from mid to far northern locations, but bright and easily observed elsewhere. (Image produced using Starry Night Pro. AllSky data courtesy of Main-Sequence Software Inc.)

NGC 1501 (Blue Oyster Nebula)

NGC 1501 is a small planetary nebula in the remarkably nondescript far northern constellation of Camelopardalis. Completely lacking in bright stars, few amateur astronomers spend much time in this constellation, but it does have a couple of worthwhile sights, one of which, Kemble's Cascade, has already been mentioned. NGC 1501 is another, and being far above the horizon at this time of year, it tends to be in the darkest part of the sky, well away from the dust and light pollution that hangs above the horizon.

NGC 1501 is less than an arcminute in diameter, and at an apparent magnitude of 13, it can easily be mistaken for a slightly blurry bluish-green star. The use of a narrowband filter or an O-III filter can be used to help clear up any confusion; although such a filter will substantially dim a star, it will have only a minimal effect on the nebula.

Once you've found NGC 1501, is there anything to see? To start with, you should be able to detect its slightly elliptical shape. Large telescopes (200 mm upwards) and high magnifications may reveal that it has slightly annular, i.e., hoop-like, shape but this isn't an easy observation to make. NGC 1501 is a little over 1 light year in diameter and about 5,000 light years from Earth.

NGC 1662 (Klingon Battlecruiser Cluster)

The Romans named the planets after their gods; the eighteenth-century French astronomer de Lacaille introduced 14 new constellations named after various bits of scientific equipment; and modern amateur astronomers have named a star cluster after a spaceship featured in a 1960s television show. So it is that the seemingly unchanging heavens reflect the changeable nature of human interests!

In fact NGC 1662 really does look remarkably like a head-on view of a Klingon battle cruiser. A straight row of bright stars about 10 arcminutes in length makes up the two wings, two shorter rows hang off each end of the first row, making the warp nacelles, and a compact clutch of stars in the middle is the bridge. It's a very neat little cluster, and a treat for any *Star Trek* fan.

NGC 1662 contains at least 20 members, is about 4 light years in diameter, and is about 1,200 light years away.

NGC 1973, 1975 and 1977 (Running Man Nebula)

The Orion Nebula is one portion of the Orion Molecular Cloud Complex illuminated by a cluster of young, hot stars; taken as a group, NGC 1973, 1975, and 1977 refer to another. In this case the emission and reflection nebula is much less strongly illuminated, so it tends to be overlooked by astronomers distracted by the Orion Nebula. That's a shame, because this reflection nebula with emission nebula components is an attractive and interesting object in its own right.

Of the three components, NGC 1973 and 1975 are the easiest to spot, being smaller and brighter. But unless your skies are reasonably dark, you might not be able to see any of them.

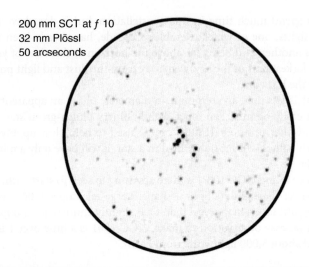

200 mm SCT at ƒ 10
32 mm Plössl
50 arcseconds

Figure 2.10. All hands to battle stations! NGC 1662 really does look like the front view of a Klingon cruiser. (Image produced using Starry Night Pro. AllSky data courtesy of Main-Sequence Software Inc.)

Try centering your telescope on the magnitude 4.6 star 42 Orionis (SAO 132230). This is one of the stars that illuminates the portion of the nebula referred to as NGC 1977, and quite a bit brighter than the nearby magnitude 7 star 45 Orionis a mere 4 arcminutes away. About 8 arcminutes away from this pair of stars is another pair of stars, quite a bit dimmer than the first pair, one of magnitude 7.3 and the other magnitude 9.7. These stars are at the centers of NGC 1973 and NGC 1975, respectively. The use of a narrowband filter or, to a lesser extent, a broadband filter will help when observing this object. Such filters will dim the stars, making them easier to see the nebulosity around them.

The nebula is 1,500 light years away and taken together a little under 10 light years in diameter.

NGC 2169 (The 37 Cluster)

This sparse open star cluster in Orion has received its common name because its members seem to be arranged in two clumps, one arranged like the number '3,' and the other like the number '7.' Fans of the works of Douglas Adams might be a bit disappointed that it isn't the number 42 that is spelled out in the night sky, but NGC 2169 is still a pretty neat arrangement of stars nonetheless.

NGC 2169 is about 3,600 light years from Earth and a mere 7 light years in diameter. It contains about 30 stars and is believed to be quite young, around 50 million years old.

NGC 2237-9, 2244 and 2246 (Rosette Nebula)

The Rosette Nebula is a big but dim emission nebula that was initially only detected piece by piece, hence its multiple NGC numbers: 2237, 2238, 2239, and 2246. Only subsequently was its shape and full extent recognized. Spanning some 1.3° of space, it is almost three times as wide as a full Moon. The nebula has a roughly circular shape and, in photographs at least, a distinctly reddish glow.

Although its apparent magnitude is comparatively high, 6.0, because this object is so big all this light is very spread out, resulting in an exceedingly dim object. Narrowband and O-III filters help significantly but can't do much about moderate to high levels of urban light pollution. You really do need dark skies to see the Rosette Nebula.

In the center of the nebula is a small but bright open start cluster, NGC 2244. Even small telescopes should reveal this cluster without problems. Together with the Rosette Nebula, the two objects make up a cloud of gas and dust around some young, hot stars a mere 3 million years in age.

Astronomers estimate that the Rosette Nebula is about 4,500 light years away and 130 light years in diameter.

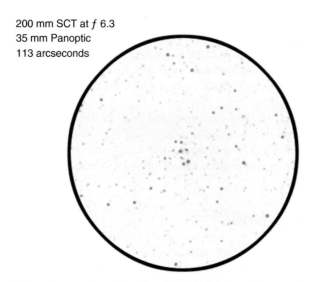

200 mm SCT at ƒ 6.3
35 mm Panoptic
113 arcseconds

Figure 2.11. The Rosette Nebula is a very difficult object to see under suburban conditions, but the open star cluster at its center, NGC 2244, is much easier to detect. (Image produced using Starry Night Pro. AllSky data courtesy of Main-Sequence Software Inc.)

NGC 2264 (Cone Nebula, Christmas Tree Cluster)

This interesting object in Monoceros is a two-for-one deal: most observers under suburban skies should be able to see the star cluster, while observers with big telescopes and dark skies get a faint nebula thrown in as well.

The Christmas Tree Cluster is, as its name suggests, an arrangement of stars that really does look a bit like a child's drawing of a Christmas tree – though admittedly an *upside-down* Christmas tree for northern observers, with the bright star at the top of cluster forming the base of the tree. Although it can be seen with binoculars, it looks best through telescopes with apertures from 100 mm upwards.

Do note that this isn't the only Christmas tree in the night sky: NGC 581 (M103) in Cassiopeia is sometimes called the Christmas Tree Cluster as well, thanks to an equally festive pattern of stars.

Surrounding the Christmas Tree Cluster is a cloud of gas and dust known as the Cone Nebula. This complex structure contains both a diffuse nebula of hot gas and a dark nebula of obscuring dust. The gas is excited by radiation from the bright star 15 Monocerotis, the star that forms the base of the Christmas Tree Cluster. 15 Monocerotis, also known as S Monocerotis, is a fascinating target in itself, and more will be said about this star in the entry for SAO 114258 in the fourth section of this chapter.

Unfortunately for the amateur astronomer, the Cone Nebula is an incredibly difficult object to see visually. Pitch black skies are essential, and a narrowband filter will be very useful for cutting out skyglow and making the nebula a bit easier to see. Even under good conditions, most astronomers only see the (relatively) bright tip of the nebula.

The Cone Nebula is in fact one of the brighter (!) parts of the nebulosity that makes up NGC 2264. At least two other bits have received names, the Snowflake Cluster and the Fox Fur Nebula, both of which are targets for astrophotographers rather than visual observers.

NGC 2264 is 2,400 light years away and believed to be between 3 and 5 million years old, making it one of the youngest open star clusters amateur astronomers can observe. Because it is both young and comparatively nearby, NGC 2264 is of particular importance to professional astronomers interested in star formation.

NGC 2323 (M50)

M50 is a fairly bright open star cluster in Monoceros about 16 arcminutes in diameter. It is quite compact, but because the background field of stars is very rich, it can be difficult to pick out. According to some, this cluster has a heart shape, but this has never been obvious to the author.

This cluster is about 20 light years in diameter and contains at least 109 stars, including a red giant (SAO 134103) and a few yellow giants. Astronomers believe that it is a relatively young cluster, around 100 million years old. M50 is almost 3,000 light years away.

NGC 2353

NGC 2353 is an open star cluster in Monoceros. It lies in a part of the Milky Way with lots of open clusters, and picking out the boundaries between them and the background stars isn't easy. With an angular width of about 20 arcminutes this is an object that looks best through a low-power, wide-field eyepiece. On a 200-mm SCT, a 32-mm Plössl does a good job, but if you can get a wider field by using either an f/6.3 reducer-corrector or a low power 2-in. eyepiece – ideally, both – then this cluster will look even better.

NGC 2353 is about 5 light years in diameter and is located about 3,400 light years from Earth. It is thought to be about 76 million years old, making it a relatively young cluster. Most of its members are hot blue stars, and the older red giants seen appear to be background stars that happen to be in the same field of view.

Stephen James O'Meara has referred to this cluster as Avery's Island, a reference to a seventeenth-century pirate by the name of Captain Avery. His idea was that the part of the sky spanning the Monoceros-Canis Major border is so filled with treasures – in this case star clusters – it would seem to be a pirate's paradise.

NGC 2360

The constellation of Canis Major is distinctive and well known for its bright star Sirius, but it is not well stocked with deep sky objects suitable for small to medium aperture telescopes. NGC 2360 is one of the few, an open star cluster about 12 arcminutes in diameter and visible even through binoculars.

This is a charming little cluster, very bright and dense. Unusually for an open star cluster, it contains some red giants as well as hot blue stars, and it is one of the older open star clusters known. Current estimates put its age at about 2 billion years, much older than most other star clusters (see the section on 'NGC 2682' for more on why this is the case).

NGC 2360 lies in a patch of sky replete with open star clusters well worth exploring with binoculars or a wide-field telescope. It is believed to be about 6,150 light years away.

NGC 2440

NGC 2440 is a small planetary nebula in the southern sky constellation of Puppis. Because of its location it is difficult to see from mid to far northerly locations. Observers in southern England for example will find that it does not rise more than about 20° above the horizon, which means that an horizon free of trees and

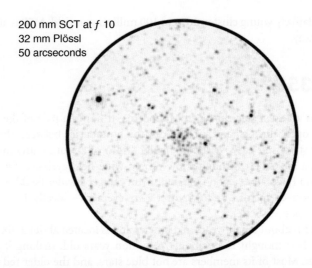

200 mm SCT at ƒ 10
32 mm Plössl
50 arcseconds

Figure 2.12. NGC 2360 is charming but often overlooked open star cluster. (Image produced using Starry Night Pro. AllSky data courtesy of Main-Sequence Software Inc.)

buildings, as well as light pollution, is required. Observers farther south will of course have an easier time of things, and for observers at the latitude of the American Midwest this is a pleasant little planetary nebula that looks good through medium to large aperture telescopes.

Because this planetary nebula is rather small, less than an arcminute in diameter, at first glance it might be mistaken for a moderately bright but out-of-focus star. The use of either a narrowband or an O-III filter will pay dividends here, making it easy to pick out the nebula from the background stars. Raising the magnification should reveal its elliptical shape, but ordinary amateur telescopes aren't big enough to show the magnitude 19 white dwarf star at its center. Oddly perhaps given its faintness, this star is apparently one of the hottest stars known to science, with a surface temperature thirty times that of the Sun!

NGC 2440 is about 3,600 light years away.

NGC 2447 (M93, Butterfly Cluster)

M93 is an attractive open star cluster in Puppis that has a reputation for being very difficult to see from mid to far northern localities because of its southerly position in the sky. A clear, dark southern horizon is essential.

A wide field is important, too, given its angular diameter of about 22 arcminutes. NGC 2477 is believed to be about 4,200 light years away and at least 10 light years in diameter. It is thought to be around 100 million years old, but unusually for such a young open star cluster it contains at least two orange supergiant stars. Both can be easily picked out with medium to large aperture telescopes.

NGC 2477

NGC 2477 is an open star cluster in Puppis with around 300 members and spanning an area about 22 arcminutes in diameter. While highly regarded by some astronomers as being a very lovely cluster, it isn't a familiar object, in part because it isn't one of the Messier objects, but surely also because it is so difficult to see. Like the other deep sky objects in Puppis, from the point of view of many northern hemisphere observers it never gets very far above the horizon. Indeed, from England it is invisible, forever hidden below the horizon.

NGC 2477 is about 4,200 light years away and thought to be around 700 million years old.

NGC 2527

This open star cluster in Puppis is a large, fairly bright object that is difficult to see from far northern locations because of its location south of the celestial equator. It's a fairly loose cluster of at least 50 stars scattered across about 22 arcminutes of space and is rather difficult to pick out from the rich background field of stars that makes up this part of the Milky Way.

NGC 2527 is about 1,920 light years away and is believed to be about 10.8 light years in width. The precise age of this cluster is somewhat uncertain but believed to be around a billion years, making this one of the older clusters in our galaxy.

NGC 2539

Another one of the nice open star clusters in the constellation of Puppis. Given the location of this constellation, to the south of the celestial equator, objects in this constellation can be difficult to see from far northern locations. But under good conditions, NGC 2539 is a surprisingly attractive non-Messier object, easily visible through binoculars and an interesting sight through any telescope.

NGC 2539 is a fairly big object, though, with an angular width of about 42 arcminutes, so a low-power, wide-field eyepiece is essential. On a 200-mm SCT, a 32-mm Plössl will deliver a 50-arcminute field of view just big enough to frame NGC 2539 against a less densely packed field of stars.

Just off to one side of the cluster is the magnitude 7.4 star 19 Puppis. Although this might appear to be part of the cluster, it isn't, and is a good deal closer. 19 Puppis is estimated to be about 185 light years from Earth, whereas NGC 2539 is thought to be about 4,445 light years away.

NGC 2548 (M48)

M48 is a large, bright open star cluster in Hydra that can be easily seen through binoculars. Under dark skies it is just visible to the naked eye.

With an angular diameter of about a degree, it is an object that looks best through a wide-field, low-power eyepiece. On a 200-mm SCT, a 32-mm Plössl will just about squeeze it into the field of view, but for a really good look at this object you'll want to use an f/6.3 reducer-corrector and a low-power 2-in. eyepiece such as a 35-mm Panoptic.

M48 contains some 80 stars spread out across about 23 light years of space. It is believed to be about 300 million years old and contains several yellow giants as well as the usual hot blue stars typical of open star clusters. It is a relatively nearby object at a mere 1,500 light years from Earth.

NGC 2682 (M67)

Compared to the celebrated Beehive Cluster M44, the other Messier object in Cancer, the open cluster M67, is far less often viewed. Despite that, it is an attractive cluster that is much easier to fit into the field of view than M44, and it's also a much richer cluster, so looks more impressive. From a scientific perspective, it also has some points of interest.

One of its quirks is its age: M67 appears to be several billion years old, its age being at least 3.2 billion years, probably closer to 4 billion years, and possibly even as high as 5 billion years. By contrast most open clusters are relatively young objects, their constituent stars drifting apart within a few hundred million years. The Pleiades, for example, are about 100 million years old, and astronomers believe that within the next 250 million years its stars will have become so separated that the Pleiades Cluster will effectively cease to exist.

How do astronomers know that M67 is so old? One clue is the large number of stars similar to our own Sun, some of which have progressed to the red giant stage. For this to be the case, the cluster will have to have been in existence for a relatively long time, implying an age measured in billions rather than millions of years.

M67 also contains thirty or so blue stragglers, stars that are hotter and bluer than stars of equivalent luminosity in the cluster. The existence of such stars is a bit of a mystery, since stars in clusters were all formed from the same cloud of gas and should evolve in more or less the same sort of way. Blue stragglers buck this trend. Precisely how these stars have managed to develop in this way isn't at all clear, but one explanation is that they have somehow absorbed material from another star. So whereas most stars are stuck with whatever stellar fuel they began life with, blue stragglers have received one or more 'top ups,' resulting in their exceptional properties.

M67 is about 2,700 light years away and believed to be about 19 light years in diameter.

NGC 3132 (Southern Ring Nebula)

This southern sky equivalent to the better known Ring Nebula (M57) in Lyra is a planetary nebula in the constellation of Vela. It's a fairly bright nebula and easy to see under even somewhat light-polluted skies. As with other planetary nebulae, the use of a narrowband or O-III filter will help confirm the identity of this object if you can't distinguish it from nearby stars.

The Southern Ring Nebula is 0.4 light years (about 3 million million km) in diameter. It is about 2,000 light years from Earth, making it one of the closest planetary nebulae to us.

Unlike the northern Ring Nebula, the Southern Ring Nebula has a relatively bright (magnitude 10) star at its center. A medium to large aperture telescope should show this central star clearly under favorably dark skies. In actual fact there are two stars at the center of this nebula, and it's the fainter of the two (a magnitude 16 star) that is actually the hot white dwarf that causes the nebula to glow.

NGC 3242 (Ghost of Jupiter Nebula)

Widely considered to be one of the best planetary nebulae in the sky, NGC 3242 is a small, bright object in the otherwise uninspiring constellation of Hydra. It is easy to spot even under somewhat light-polluted conditions, and the use of a narrowband or O-III filter to darken the background sky and adjacent stars should remove any doubt about its identification.

Unusually for a deep sky object, this planetary nebula has a certain amount of color, most observers reporting that it appears to be blue or blue–green.

Its common name, the Ghost of Jupiter, is a reference to its apparent size at the eyepiece, 25 arcseconds, slightly less than that of Jupiter, which ranges from 30 to 50 arcseconds. Just as you'd do when viewing Solar System objects, try using quite high levels of magnification so that its details become more apparent. Using magnifications of up to ×800, noted deep sky observer David Knisely at the Prairie Astronomy Club in Lincoln, Nebraska, likens the nebula to the eye-shaped logo of the Columbia Broadcasting Company (CBS). What do you see?

Of course, although Jupiter and the Ghost of Jupiter have about the same size when viewed through a telescope, the two objects are very different in absolute size: whereas the planet Jupiter has a diameter of about 143,000 km, the Ghost of Jupiter Nebula is about 20 million times larger, measuring a whopping 2,842,362,000,000 in diameter. But because the Ghost of Jupiter Nebula is also a lot farther away, about 1,400 light years away, compared to a mere 33 light minutes in the case of Jupiter, the two objects seem to be of similar size. Although these numbers may seem – literally as well as figuratively – astronomical, they do give you some idea of the scale of the universe.

The only downside to this object is that for far northern observers it may not rise very far above the southern horizon. In southern England, for example, it gets to about 20° or so above the horizon. Although your narrowband or O-III filter may help deal with ambient light pollution, they can't do anything about inconveniently placed trees or houses!

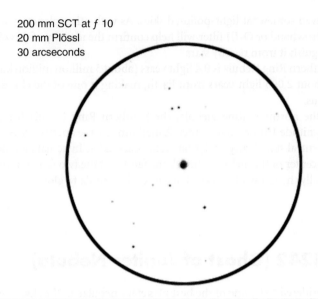

200 mm SCT at ƒ 10
20 mm Plössl
30 arcseconds

Figure 2.13. NGC 3242 is a small, disc-shaped planetary nebula. A narrowband or O-III filter is a useful tool when viewing this object. (Image produced using Starry Night Pro. AllSky data courtesy of Main-Sequence Software Inc.)

NGC 3228

NGC 3228 is a small but bright open star cluster in Vela with an angular diameter of about 5 arcminutes. Although easily seen through binoculars under dark skies, this attractive cluster of mostly hot, blue stars is very much a southern sky object, and difficult to see even from latitudes as low as central Florida.

NGC 3228 is about 1,600 light years from Earth and is estimated to be a relatively young cluster around 42 million years in age. Stephen James O'Meara calls this cluster the Queen's Cache Cluster, but whether this name catches on remains to be seen.

NGC 4590 (M68)

Although it may be the biggest of all 88 constellations, Hydra is singularly lacking in bright deep sky objects. One of the few is M68, a reasonably bright globular cluster easily visible through binoculars under good conditions and a lovely sight when viewed with mid to large aperture telescopes. It is rated at Class X on the Shapley–Sawyer scale.

As its Shapley–Sawyer class rating suggests, M68 is a fairly loose globular cluster, but the core is impossible to resolve visually even in large aperture telescopes. Unfortunately for mid to far northern observers, it's a relatively low-lying object that doesn't get very far above the southern horizon, which is perhaps one reason why it is one of the less familiar Messier Catalog globular clusters.

M68 is about 106 light years in diameter and lies about 33,000 light years from Earth.

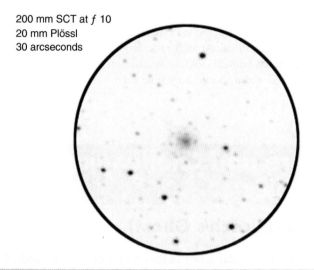

200 mm SCT at ƒ 10
20 mm Plössl
30 arcseconds

Figure 2.14. M68 is a small, fairly loose globular cluster that lies in a relatively southerly part of the huge constellation Hydra. (Image produced using Starry Night Pro. AllSky data courtesy of Main-Sequence Software Inc.)

NGC 7662 (Blue Snowball Nebula)

This small planetary nebula is sufficiently bright that it can be spotted through any telescope, but you do need reasonably dark skies and an aperture upwards of 200 mm to see its famous blue color. Like most planetary nebulae, its small angular diameter, a little over half an arcminute, means that it can easily be confused with an out-of-focus star. You will need quite high levels of magnification to see any details. Given sufficient magnification, a fainter halo around the bright central disc should be apparent. Narrowband and O-III filters are well worth using.

The precise distance between NGC 7662 and Earth is a matter of some debate, the usual value being placed at around 3,200 light years but some astronomers believing the distance could be nearly twice that. Since estimates of its diameter depend on how far away it is, there's some uncertainty in this regard as well, but the minimum value is likely around 0.3 light years, just under 3 million million km. If NGC 7662 is farther away that had been previously assumed, then it would have to be larger as well.

NGC 7686

NGC 7686 is an open cluster in the constellation of Andromeda. It is about 13 light years in diameter and estimated to be about 3,000 light years from Earth.

It is reasonably bright and compact, with an angular diameter of about 15 arcminutes, and should be framed very nicely in the field of view a low-power eyepiece such as a 32-mm Plössl. At least nine bright stars should be visible, together with many more much fainter stars. The bright stars are at least between about sixth and tenth magnitudes and should be visible even under moderately light polluted skies.

Obscure and Challenging Deep Sky Objects

NGC 404 (Mirach's Ghost)

NGC 404 is a dwarf lenticular galaxy in the constellation Andromeda, 6 arcminutes from the star Beta Andromedae (sometimes called Mirach). Because of its proximity to this star it is very difficult to see or even photograph, the glare from the star swamping out the light from this relatively dim galaxy. As a result it has become known as Mirach's Ghost.

At the eyepiece this galaxy can seem very small and condensed, almost star-like. It can be seen with telescopes from 100 mm upwards, but it is easily overlooked in the glare of Mirach or mistaken for an out-of-focus star. Moving the star Mirach out of the field of view is the key, either by switching to a higher power or narrower field-of-view eyepiece, or else by slewing the telescope very slight off to one side.

NGC 404 is about 10 million light years away and 65,000 light years in diameter.

NGC 891

NGC 891 is a spiral galaxy in Andromeda that is viewed edge-on, but because it has a low surface brightness, it is dim and difficult to see. A large aperture telescope is required, as well as fairly dark skies. Even then, a broadband filter can be useful for blocking skyglow and any slight light pollution.

NGC 891 is about 30 million light years away and part of the Coma-Sculptor Cloud of galaxies. It is believed to be about 110,000 light years in diameter, making it comparable to our own Milky Way Galaxy in size as well as shape.

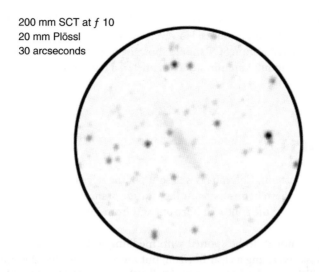

200 mm SCT at ƒ 10
20 mm Plössl
30 arcseconds

Figure 2.15. Spotting NGC 891 under suburban sky conditions will be difficult. (Image produced using Starry Night Pro. AllSky data courtesy of Main-Sequence Software Inc.)

NGC 1232

This spiral galaxy in the constellation of Eridanus is viewed face-on, and, as is often the case with galaxies in this orientation, its surface brightness is much lower than its visual magnitude would suggest. Indeed, this galaxy is very difficult to see under anything other than dark sky conditions. An aperture of 200 mm is certainly required under most conditions, and the best views come with much larger telescopes.

NGC 1232 is about 70 million light years away and part of the Eridanus Cluster of galaxies. It has a close neighbor, the galaxy NGC 1232A. The gravitational pull of this satellite galaxy has distorted the spiral arms of NGC 1232 in a way similar to that seen with the Whirlpool Galaxy (M51).

NGC 1788

NGC 1788 is a reflection nebula in Orion. It is small and notoriously difficult to see except from dark sky locations. The problem is that reflection nebulae shine by reflecting starlight from nearby stars, in this case a cluster of young stars actually hidden from view behind the nebula. Because reflection nebulae aren't only emitting the wavelengths of light that narrowband and O-III filters are designed to let through, neither of these filters dramatically improves the view. A standard light-pollution reduction filter might be even more helpful if street lighting is the prime source of light pollution in your area.

From dark sky environments NGC 1788 can be seen with relatively small aperture telescopes, from 100 mm upwards. However, suburban astronomers will want to use larger telescopes. A 200-mm SCT should show the brightest region of the nebula, but since this is only about 3 arcminutes in diameter, a fair amount of magnification is required.

NGC 1851

This globular cluster in Columba is well south of the celestial equator and difficult to see from the northern hemisphere. At the latitude of the midwestern United States it barely gets 9° above the southern horizon, and from England it is always below the horizon.

It is bright enough to be spotted with medium aperture telescopes under dark sky conditions, assuming a clear horizon, but more realistically a 200-mm aperture telescope will be required to see any detail. It's reasonably bright and quite large (about 18 arcminutes) and an attractive object through large telescopes and under at least somewhat dark skies.

NGC 1851 is about 40,000 light years away and 140 light years in diameter. Like most other globular clusters, it is an ancient object, around 14 billion years old.

NGC 1904 (M79)

The constellation of Lepus contains very few deep sky objects easily viewed through telescopes of the types owned by amateur astronomers. There's really only a single deep sky object of note, the globular cluster M79. It is a fairly dim object, though, and isn't an easy target for small to medium aperture telescopes under suburban skies, though at magnitude 7.7 it can be detected with binoculars under pitch black skies. Lepus is a relatively southerly constellation, and M79 is difficult to see from far northern locations. Observers in southern England will find it only rises about 13° above the horizon.

M79 is an odd globular cluster; it is unlike most other globular clusters in being best viewed in winter rather than summer. Most globular clusters orbit the galactic core, so we tend to view them by looking into that part of the Milky Way. M79 is different and actually orbits a long way out, 60,000 light years from the core.

The origins of M79 are uncertain, but one explanation is that it belongs to the Canis Major Dwarf Galaxy, a nearby galaxy being pulled apart by the Milky Way Galaxy. As this interaction progresses, bits of the Canis Major Dwarf Galaxy are absorbed by the Milky Way Galaxy, including M79. Although this explanation isn't universally accepted by astronomers, it does at least explain why M79 is where it is.

M79 is about 42,000 light years from Earth and is thought to be about 118 light years in diameter.

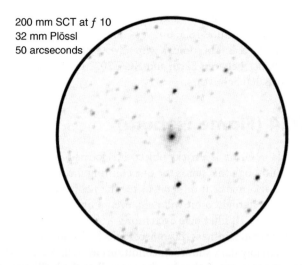

200 mm SCT at *f* 10
32 mm Plössl
50 arcseconds

Figure 2.16. M79 is one of relatively few globular clusters visible during the winter months. It is rather faint, though, and its southerly location makes it a challenging object for far northern observers. (Image produced using Starry Night Pro. AllSky data courtesy of Main-Sequence Software Inc.)

NGC 1981

NGC 1981 is an open cluster at the northern end of the Sword of Orion. It is sparse and spread out across 25 arcminutes and looks best at low magnifications. The brightest of its few stars are arranged in a distinctive Y-shaped asterism. This cluster is thought to be about 1,500 light years away.

NGC 2022

Although Orion is famous for its emission nebulae, it is not a constellation most astronomers think of when talking about planetary nebulae. That's perhaps why the planetary nebula NGC 2022 is so often overlooked.

In general terms it is a very ordinary nebula. Through a 200-mm SCT it appears as a small (18 arcsecond diameter) elliptical blur. Unlike the showpiece planetary nebulae, is not obviously colored, and simply looks like a faint gray spot. It can be easily mistaken for an out-of-focus star, but unlike stars, it is not substantially dimmed by a narrowband or O-III filter.

Beyond its general shape, there isn't any obvious detail, even at high magnifications. Long-exposure photographs reveal a two-part structure, a bright inner core (the bit we can see with amateur astronomy telescopes), and a fainter shell of gas around the outside (only visible in photos or through large research-grade observatory telescopes). The outer shell of gas extends across a diameter of about 88 arcseconds, something like 3 light years of space. The star in the center of the nebula is very dim, around magnitude 16.5, but very hot and seemingly close to the point where it is cooling down as it becomes a white dwarf star.

The precise distance between Earth and NGC 2022 is not known, but is believed to be about 6,000–8,000 light years.

NGC 2024 (Flame Nebula)

The Flame Nebula in Orion is a tricky object. It is located rather close to the bright star Alnitak, and although this star is the one emitting the light that makes the nebula glow, its glare also makes it difficult to see the nebula at all. Almost any filter that tones down the starlight a bit will help, even a light pollution filter, though a narrowband filter or O-III filter help even more.

Between of its overall faintness this object looks best under dark skies, but it can be seen from reasonably dark suburban conditions as well. A narrowband or O-III filter will help cut out some of the sky glow as well as the brilliance of Alnitak, so they're doubly useful in this regard. Through a 200-mm SCT the Flame Nebula is visible as a faint patch of light with a dark band, a dust cloud, running down the middle.

Darker skies and larger telescopes reveal more detail, including more dust clouds, and also show more of the nebula. Under optimal conditions, it can be seen to be remarkably large, about 30 arcminutes in diameter, though suburban astronomers will usually only see the bright central region.

The Flame Nebula is part of the Orion Molecular Cloud Complex mentioned in the section on the Orion Nebula (NGC 1976 and 1982).

NGC 2194

This open cluster is rather peculiar. Unlike most open clusters, it contains mostly metal-poor stars. These are the types of stars commonly found in globular clusters. Most open clusters tend to have metal-rich stars, stars rather like our own Sun. Part of the reason for this is that, unlike most open clusters, this cluster is far from the center of the galaxy (and some 10,000–12,000 light years from Earth). It is also an old star cluster, around 550 million years old in fact, making it much older than most open star clusters.

At the eyepiece this nebula is rather faint and may be difficult to see with small telescopes. Medium to large aperture telescopes reveal an attractive mix of blue and yellow stars.

NGC 2232

This bright open cluster includes the bright star 10 Monocerotis, so it should be easy to pick out from the background field of stars. It is a loose cluster, spread out across some 29 arcminutes of space; thus a low-magnification, wide-field eyepiece is required for the best view. It contains over 40 stars, is 1,200 light years away, and thought to be about 40 million years old.

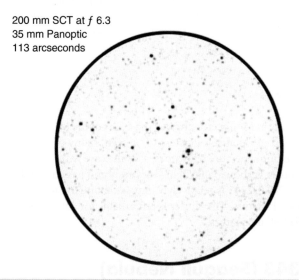

200 mm SCT at ƒ 6.3
35 mm Panoptic
113 arcseconds

Figure 2.17. NGC 2232 is best appreciated with a wide field, low power eyepiece. (Image produced using Starry Night Pro. AllSky data courtesy of Main-Sequence Software Inc.)

NGC 2251

NGC 2251 is one of numerous open clusters in Monoceros. It is easily seen through telescopes of all apertures as a tight cluster of stars surrounding a bright (ninth-magnitude) star in the center. NGC 2251 is about 4,300 light years away and is 6 light years in diameter.

NGC 2261 (Hubble's Variable Nebula)

This object's common name is unusually informative: it is indeed a nebula, it was discovered by Edwin Hubble, and its brightness varies across an irregular period of several weeks.

Under dark sky conditions the brightest portion of this combination reflection and emission nebula can be spotted with quite small telescopes. A telescope with an aperture of 100 mm should show the bright portion at the apex of this comma-shaped nebula, while larger telescopes allow more of its shape to become apparent. It is inevitably the case that suburban astronomers will have difficulty seeing the fainter parts of the nebula, but reasonably dark suburban skies should allow a 200-mm SCT to show the brightest part of this nebula.

Because this object is largely a reflection nebula, light-pollution filters are of limited value. If you must use one, a broadband filter is probably the best bet.

What makes the Variable Nebula variable? At the apex of the nebula is an irregular variable star, R Monocerotis. As its brightness varies, so, too, does the brightness of the nebula around it.

NGC 2301

Observers looking at Monoceros are viewing a part of the Milky Way thick with stars. There are numerous open star clusters to be seen, many of which are good targets for small to medium aperture telescopes, as is the case with NGC 2301.

This star cluster is about 3,000 light years away and has an angular width of about 12 arcminutes. NGC 2301 is relatively young, at about 164 million years of age, and contains at least 146 stars. The cluster is remarkable for sitting very close to both the galactic and celestial equators.

NGC 2343 (Seagull Nebula)

NGC 2343 is an open cluster in Monoceros that is easily viewed through small telescopes. However, through the eyepiece it isn't very large, about 6 arcminutes across, and can be easily overlooked against the background stars. If you look closely, though, you'll notice a group of mostly blue stars with one fairly bright orange star off to one side; this is NGC 2343. It is believed to be about 4,300 light years away and around 130 million years of age.

NGC 2343 and another nearby cluster, NGC 2335, are both part of a large region consisting of star clusters, clouds of dust, emission nebulae, and reflection nebulae. The whole system is sometimes called the Seagull Nebula, though amateur astronomers typically apply that name to the relatively bright southerly region of the complex cataloged as IC 2177. It doesn't rise above the horizon at mid to far northern latitudes and is essentially a southern hemisphere target only.

Even at latitudes where the object is favorably placed, it is very dim and difficult to see, and requires a large aperture telescope and relatively dark skies. Light pollution filters help only somewhat, with narrowband and H-beta filters offering the most improvement.

NGC 2359 (Thor's Helmet)

This object consists of an emission nebula surrounding a peculiarly hot type of supermassive star called a Wolf-Rayet star. Wolf-Rayet stars are approaching the point at which they become supernovae. The nebula is notoriously difficult to see visually, though light pollution filters, in particular O-III filters, can help significantly. Dark skies are almost a prerequisite, and while the nebula has been detected with medium aperture telescopes, large aperture telescopes are usually required.

NGC 2359 is at least 15,000 light years away and has a width of about 30 light years. At the eyepiece it is often a lot smaller than people expect when looking at photographs, having an angular diameter of only 10 arcminutes.

NGC 2362

The forty or so hot blue stars that make up this attractive open star cluster in Canis Major seem to be swarming around the bright star Tau Canis Majoris. In fact this star may not belong to the cluster at all and might just be in our line of sight, though this is still a matter of debate.

In any case, NGC 2362 is an easy object for telescopes of all sizes, but it looks really nice in telescopes of apertures 150 mm upwards. With an angular diameter of about 8 arcminutes it is rather small and benefits from a bit of magnification, though it can be difficult to pick out from the rest of the star field. Through a 200-mm SCT and a 20-mm Plössl it looks particularly nice.

NGC 2362 is a young cluster, perhaps 5 million years old, and long-exposure photographs reveal traces of gas and dust around its stars. Indeed, some of this material seems to be what astronomers call protoplanetary discs. These are the discs of gas and dust around young stars that eventually condenses to form planets, asteroids, comets, and other Solar System objects. The cluster is about 4,000 light years away.

NGC 2371 and 2372

This odd planetary nebula has two bright regions, and initially each region was counted as a separate nebula, hence the two NGC designations. Appropriately enough for a single deep sky object with a dual identity, NGC 2371 and 2372 are to be found in the constellation of Gemini.

A telescope of 200 mm is enough to show the two bright spots on either side of this planetary nebula, assuming reasonably dark skies. However, its small angular diameter – a bit over 2 arcminutes in width and about 1 arcminute in height – means that it can be difficult to distinguish from neighboring stars.

Use either a narrowband filter or an O-III filter to double check that you are looking at the right object: such filters dim stars greatly, but planetary nebulae hardly at all. If your telescope has sufficient aperture, use of such filters make also make it easier to see the fainter parts of the nebula that surround and connect the two bright spots. Doing this is challenging, though, under suburban skies.

The whole nebula is about 4,400 light years away.

NGC 2403

NGC 2403 is a spiral galaxy in Camelopardalis. It is fairly bright, and under good dark skies can be seen through binoculars. But suburban astronomers will need reasonably large apertures to be sure of bagging this galaxy. Telescopes above 150 mm should show its bright core easily enough, but you will need much larger telescopes, 250 mm upwards, to see the fainter regions around the core.

NGC 2403 is part of the comparatively nearby M81 group and is believed to be about 8 million light years away. It is believed to be about 21,900 light years in diameter.

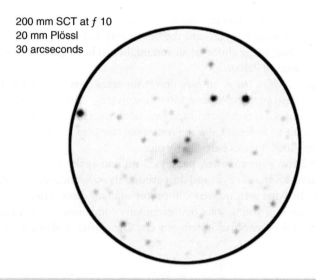

200 mm SCT at ƒ 10
20 mm Plössl
30 arcseconds

Figure 2.18. NGC 2403 is quite a large galaxy, but its surface brightness is low. (Image produced using Starry Night Pro. AllSky data courtesy of Main-Sequence Software Inc.)

NGC 2419 (Intergalactic Tramp)

This interesting but notoriously dim globular cluster is very difficult to find without a go-to telescope, lying in a part of the sky singularly lacking in bright stars. Although owners of go-to telescopes will at least be able to get NGC 2419 into the field of view easily enough, actually seeing the thing is a whole other problem! Some amateurs have reported being able to see NGC 2419 through telescopes with apertures as small as 100 mm under very dark skies, but from a reasonably dark suburban location, it is a challenge even with a 200-mm telescope. Once you have spotted NGC 2419, what you see is more like an elliptical galaxy than a globular cluster, the object appearing as little more than a faint circular blur less than 9 arcminutes in diameter.

So why bother? NGC 2419 is worth looking at because it is so distant. It orbits the galactic core at a distance of 300,000 light years, a good 275,000 light years from Earth. To put that into perspective, Omega Centauri is a mere 15,000 light years from us, and the Great Hercules Cluster only 25,000 light years away. So although apparently similar to Omega Centauri in terms of size and actual luminosity, from our earthbound point of view it seems incredibly small and faint.

For a long time this globular cluster was known as the Intergalactic Tramp or the Intergalactic Wanderer, the prevailing belief being that NGC 2419 was too far away from the galactic core to be held in place by gravity, and that it was instead a globular cluster that drifted through intergalactic space all by itself. That idea is not now generally accepted, but NGC 2419 remains an odd creature, farther out from the galactic core than the Milky Way's two satellite galaxies, the Magellanic Clouds. How did NGC 2419 get so far out? Was it formed alongside the Milky Way Galaxy or somewhere else? One explanation is that NGC 2419 is the core of a dwarf galaxy that drifted by the Milky Way Galaxy at some point in time.

NGC 2506

NGC 2506 is one of many open clusters in Monoceros. It is quite small and easily overlooked, but an attractive enough object when viewed through medium to large aperture telescopes. It is only 7 arcminutes in diameter, though, so a certain amount of magnification is helpful. On a 200-mm SCT, a 20-mm Plössl frames the cluster nicely.

This cluster is over 11,000 light years away and apparently very old for an open cluster, current estimates suggesting an age of at least 1.5 billion years.

NGC 2547

This superb open cluster in the southern sky constellation of Vela cannot be seen by observers at mid to high northern latitudes. Even in central Florida it does not rise more than about 13° above the horizon.

NGC 2547 is quite broad, around 36 arcminutes in diameter, and contains at least 50 stars. It is about 2,000 light years from Earth and about 15 light years in width.

NGC 2571

NGC 2571 is an open cluster in the southern sky constellation of Puppis and difficult or impossible to see from mid to high northern locations. But from locations where it can be seen, such as Florida, it's a compact, fairly bright open cluster that looks like a hazy patch through binoculars but can easily be resolved to its individual stars with small telescopes 100 mm in aperture upwards.

NGC 2571 is about 4,500 light years away and believed to be about 50 million years old. It appears to contain a white dwarf star. Since open clusters usually contain quite young stars, the presence of a white dwarf is rather unusual.

NGC 2655

This lenticular galaxy in Camelopardalis is not difficult to see, and its bright core can be seen with binoculars under pitch black skies. It's an easy galaxy for telescopes 100 mm upwards, though not much will be seen beyond the core, and really large telescopes and good dark skies are needed to make out the fainter region around the core.

NGC 2655 is about 80 million light years away and the brightest member of its own group of galaxies, the NGC 2655 Group, within the Virgo Supercluster.

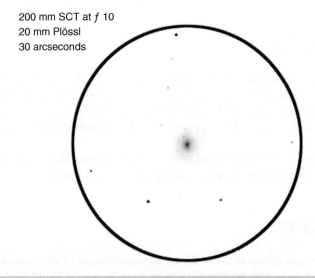

200 mm SCT at ƒ 10
20 mm Plössl
30 arcseconds

Figure 2.19. NGC 2655 is a lenticular galaxy with a fairly bright core. (Image produced using Starry Night Pro. AllSky data courtesy of Main-Sequence Software Inc.)

NGC 3003

NGC 3003 is a small spiral galaxy in Leo Minor about 80 million light years away. It is very difficult to see this galaxy with small telescopes, and even apertures of 200 mm only reveal this galaxy under reasonably dark skies.

Our view of NGC 3003 is edge-on, so we don't really see very much, but assuming your skies are reasonably dark, the central core should be apparent. Darker skies and larger apertures will show more of the shape of the galaxy, but details such as mottling or dust lanes are not apparent.

NGC 3293 (Little Jewel Box Cluster)

NGC 3293 is another deep sky object only visible to northern observers at very low latitudes. It is a very pretty open cluster in the constellation of Carina with numerous hot blue stars and several red giants. Although not quite in the same league as the real Jewel Box Cluster (NGC 4755) in Crux, it's certainly a showpiece object at latitudes where it can be seen.

NGC 3293 is 8,400 light years away and is believed to be a mere 10–20 million years old, making it a very young object by open star cluster standards.

NGC 3344

Leo Minor is one the smallest and dimmest constellations in the sky, with no stars brighter than fourth magnitude. It's of little interest to amateur astronomers, even though it does lie in a part of the sky rich with galaxies. Most of those galaxies are far too faint to be seen with the average amateur astronomer's telescope, though there are a few just about worth looking for, of which NGC 3344 is perhaps the best.

NGC 3344 is a reasonably bright galaxy only 20 million light years away, and a member of a group of galaxies referred to as the Leo Spur. This is a region of the Virgo Supercluster close to the Local Group within which own galaxy belongs, so while Leo Spur galaxies are farther away than Local Group galaxies such as the Andromeda Galaxy, they're still comparatively nearby.

NGC 3432

NGC 3432 is a spiral galaxy in the small constellation Leo Minor. It is small and faint, and consequently difficult to see. You'll need at least 200 mm of aperture and fairly dark skies to spot this edge-on galaxy, in which case all you'll see is a dim, thin streak anyway.

NGC 3432 is part of the Leo Spur group of galaxies and is believed to be about 48 million light years away.

NGC 3621

This elliptical galaxy in Hydra is somewhat faint, but under good dark skies its core, at least, is easily visible with medium to large aperture telescopes, from around 100 mm upwards. It is too far south for most northern observers to get a good look at, though, and even from central Florida it doesn't quite reach 30°

above the horizon. Consequently it is not a well known galaxy among amateur astronomers.

NGC 3621 is about 20 million light years away and part of the Leo Spur group of galaxies. It is a fairly large galaxy with a diameter of at least 93,000 light years.

NGC 5128 (Centaurus A)

Although a dramatic and easy object for observers at equatorial and southern latitudes, observers at anything other than the lowest northern latitudes will find Centaurus A an impossible target. In central Florida, for example, it rises a little higher in the sky than Omega Centauri (NGC 5139) but has a much lower surface brightness, and is consequently far more difficult to see. So although it's always worth having a stab at, don't be too disheartened if you can't see this object.

Under good conditions, this object is bright and easy to see even with a small telescope. A large telescope, anything from 150 mm upwards, should show the prominent dust lane across the middle of the galaxy. Technically a lenticular galaxy, this dust lane is unusually well developed for this type of galaxy, and astronomers believe that Centaurus A was formed by a collision between a big elliptical galaxy and a smaller spiral galaxy. Another quirk is its incredibly strong emission of radio waves. These are believed to come from a huge black hole at the center of the galaxy, a black hole estimated to have a mass equal to one billion times that of our Sun.

Centaurus A is a member of the Centaurus A/M83 Group of galaxies that, as its name suggests, is subdivided into two smaller groupings, one centered around Centaurus A and the other around Messier 83. Centaurus A itself is about 14 million light years away.

NGC 5139 (Omega Centauri)

Like Centaurus A, Omega Centauri is a southern sky object, but it is surprisingly easy to see from the latitude of central Florida. To be fair, your view won't be anywhere near as good as that enjoyed by a southern sky observer, but it's still a pleasingly large and bright object.

Under favorable conditions Omega Centauri can be seen with the naked eye, and southern sky observers consider this object to be one of the two best globular clusters in the sky, the other being 47 Tucanae (NGC 104). Poor old M13 (NGC 6205) in Hercules is very much an also-ran in comparison! Even with a low-power eyepiece, the sheer size of Omega Centauri is impressive, having an angular diameter of a whopping 36 arcminutes, making it appear bigger than the full Moon!

Omega Centauri is about 16,000 light years away. It is unlike most other globular clusters in that it contains multiple generations of stars rather than Population II stars only. One explanation is that Omega Centauri is actually the core of a dwarf galaxy that drifted too close to the Milky Way; most of its stars were absorbed into the Milky Way Galaxy, but the core remains as a globular cluster.

NGC 5236 (M83, Southern Pinwheel Galaxy)

M83 is a barred spiral galaxy in Hydra, and at only 15 million light years away, one of the closest of its type to Earth. Unfortunately for northern hemisphere observers this southern sky object is not easy to see. Indeed, mid to far northern hemisphere observers may not be able to see it at all: in southern England it only gets to about 8° above the southern horizon. M83 is the southernmost galaxy in the Messier Catalog and consequently one of the most difficult objects on the list for northern hemisphere observers to see.

This galaxy is viewed face-on, and as is often the case with galaxies oriented this way, it can be difficult to pick out from the background sky. Anything worse than the slightest amount of light pollution will obscure the galaxy completely, so dark or at least rural skies are important.

Five supernovae have been observed in M83, the first in 1923 and the most recent in 1983. It is part of its own cluster of galaxies, the M83 Group, a group that includes at least 14 galaxies in total, including Centaurus A (NGC 5128).

Colorful and Curious Stars

SAO 12969 (Kemble's Cascade) – See also NGC 1502

Kemble's Cascade is a famous asterism in Camelopardalis and is a remarkably straight chain of more than twenty stars. The chain runs for over 2.5° in length and as such is a showpiece target for astronomers using binoculars and rich-field telescopes. It is named after Fr. Lucian Kemble, the amateur astronomer and Franciscan friar who discovered the asterism.

Kemble's Cascade cannot be seen well through a standard 200-mm SCT. At best, a combination of a reducer-corrector and a 2-in. eyepiece such as a 35-mm Panoptic will let you see about three-fourths of the chain. Have the telescope locate SAO 12969, the brightest star in the chain and conveniently located close to its center. Although you still won't see the whole of Kemble's Cascade, you'll see enough to be impressed. If your telescope is equipped with a 50 mm finder, be sure to take a look at Kemble's Cascade through it. You may decide the view looks better through the finder than through the telescope!

Kemble's Cascade isn't a star cluster, though it might appear to be one. It's merely a chance alignment of stars. There are countless others such alignments, some with names, most without.

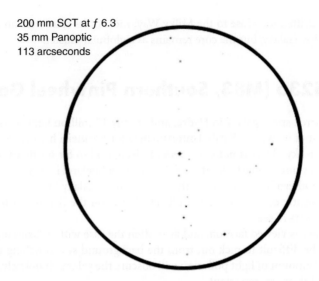

200 mm SCT at ƒ 6.3
35 mm Panoptic
113 arcseconds

Figure 2.20. SAO 12969 is the star at the center of Kemble's Cascade. (Image produced using Starry Night Pro. AllSky data courtesy of Main-Sequence Software Inc.)

SAO 23906 (Stock 23)

SAO 23922 is a magnitude 7.5 star in a fairly large and loose open cluster known as Stock 23 in Camelopardalis. This cluster is about 25 arcminutes in diameter, so a reasonably large field of view is important. Within the cluster is a distinctive keystone-shaped asterism. Most of the stars are white, but there are at least two obviously yellow ones, including SAO 23906. Stock 23 is about 30 million years old and 1,240 light years away.

SAO 25939 (12 Lyncis)

12 Lyncis is a classic triple star, but it is often overlooked because it is located in the infamously dim constellation of Lynx, about which it has been said you need the eyes of a lynx simply to find it. Owners of go-to telescopes are saved the chore of finding their way around this constellation and can instead direct their attentions to enjoying this attractive triple star. At medium magnifications a blue–white magnitude 5.0 star will be seen with a yellow magnitude 7.2 companion. With higher magnification a medium aperture telescope will reveal that the brighter star is actually two stars very close together, one about magnitude 5.4 and the other magnitude 6.0.

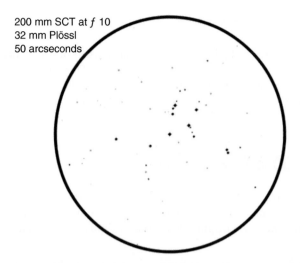

200 mm SCT at *f* 10
32 mm Plössl
50 arcseconds

Figure 2.21. SAO 23906 is the yellow star at center of the large, loose cluster known as Stock 23. (Image produced using Starry Night Pro. AllSky data courtesy of Main-Sequence Software Inc.)

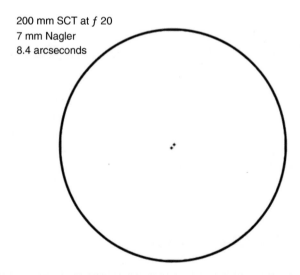

200 mm SCT at *f* 20
7 mm Nagler
8.4 arcseconds

Figure 2.22. With additional magnification, the dimmer of the two components that make up 12 Lyncis will be revealed to be a double star in its own right. (Image produced using Starry Night Pro. AllSky data courtesy of Main-Sequence Software Inc.)

SAO 26051 (15 Lyncis)

15 Lyncis is an attractive double star in Lynx that consists of two golden yellow stars, one at magnitude 4.7 and the other magnitude 5.8. They are only 0.9 arcseconds apart, and high magnifications will be required, as well as a medium aperture or larger telescope.

SAO 26312 (19 Lyncis)

This is a good triple star for telescopes of all sizes. The two brightest components are blue-white stars at magnitudes 5.8 and 6.9, separated by just under 15 arcseconds. The third component is a magnitude 7.6 star of similar color, about 3.5 arcminutes away.

SAO 55347 (Iota Trianguli)

Iota Trianguli, often referred to by its Flamsteed designation, 6 Trianguli, is an attractive double star consisting of a yellow magnitude 5.3 primary and its vaguely bluish magnitude 6.9 companion. They are separated by 3.9 arcseconds and are easily split with even small aperture telescopes given steady skies and sufficient magnification.

SAO 60198 (Alpha Geminorum, Castor)

Castor is one of the classic double stars of the sky. The two components are similar in colour, white, and also similar in brightness, one at magnitude 1.9 and the other at magnitude 2.2. They are quite close, a mere 2.2 arcseconds apart, but this isn't as problematic as it might seem. Double stars with two equally bright components are the easiest to split, the degree of difficulty increasing the more dissimilar the two stars are in brightness. Castor fits the pattern neatly and can be split with very small apertures, as little as 75 mm, given steady skies and sufficient magnification.

SAO 61391 (38 Lyncis)

38 Lyncis is a pretty double star that requires at least medium aperture telescopes to split. The magnitude 3.9 primary is blue–white and the magnitude 6.1 secondary is also blue-white, though some observers report it as being lilac in color! The two components are separated by 2.6 arcseconds.

SAO 75020 (Struve 183)

This triple star system in Triangulum is a challenging target for medium to large aperture telescopes. The primary pair AB consists of a yellow magnitude 7.7 star and a blue–white 8.4 star just 0.6 arcseconds apart. The C component is much easier to see, a magnitude 8.7 star 5.6 arcseconds away from the AB pair.

SAO 80416 (Iota Cancri)

This double star looks like a somewhat dim, washed out version of Albireo, but it is easily split with even a small telescope, and consequently a popular addition to wintertime observing schedules. The primary (SAO 80416) is a yellow giant star at magnitude 4.0, while the secondary (SAO 80415) is an A-type blue-white main sequence star at magnitude 6.6.

SAO 96265 (38 Geminorum)

38 Geminorum is a nice double star consisting of a white primary and a yellow secondary, shining at magnitudes 4.7 and 7.7, respectively. The two stars are separated by about 7 arcseconds and are easily split even with relatively small aperture telescopes.

SAO 112740 (Gamma Orionis, Bellatrix)

Bellatrix is one of the hotter blue stars in the sky, and a classic example of a B-type blue giant star. It has a surface temperature of around 21,000 K and a distinctive steel-blue coloration.

SAO 112921 (Lambda Orionis, Meissa)

Meissa is a fairly easy wintertime double star for medium and large aperture telescopes. It consists of a magnitude 3.5 primary and a magnitude 5.5 secondary, the two stars 4.3 arcseconds apart. Both stars are essentially white in color.

SAO 113271 (Alpha Orionis, Betelgeuse)

Betelgeuse is a red supergiant, and one of the biggest stars known. It is over 900 times the diameter of the Sun, and more than 18 times its mass. A popular way of describing its size is to consider what would happen if the Sun was the same size as Betelgeuse: the outer edge of the Sun would be somewhere between the orbits of Mars and Jupiter!

Having said this, because Betelgeuse is so big, its mass is spread out incredibly thinly, and most of the star would be about as matter-rich as a laboratory-grade vacuum. It's difficult to imagine what a starship pilot approaching Betelgeuse would see, but it is certain that Betelgeuse would not have sharp, clear edges to its disc of the type we see around the Sun when it is observed (through an appropriate filter, of course). Presumably it would be like a huge orangey mass, tenuous at its edges, but becoming richer towards its center.

Betelgeuse is a an irregularly variable star, though its precise brightness is difficult to estimate by eye because of the distance between it and appropriately bright stars of similar color. Roughly speaking it varies between magnitudes 0 and 1, though the exact numbers are debatable, but likely around 0.3 and 1.2. In any event, its period is a long one, around 6–7 years, but given to considerable variation.

Because Betelgeuse is so large, it is exceptional among stars in that its angular diameter can actually be measured with some degree of accuracy. By doing this, astronomers have discovered that its size actually varies, and does so along with variations in its brightness. Betelgeuse has also been photographed, and these pictures show variations in brightness across its surface – what on the Sun would be called sunspots.

The fate of Betelgeuse is interesting. It is a very massive star, and because of this, it appears to be consuming its nuclear fuel at an incredible rate. Given its size and type, it is a comparatively old star, some 8.5 million years in age, and scientists believe its final collapse into a supernova is likely to happen in the relatively near future.

SAO 114146 (Plaskett's Star)

Plaskett's Star is a visually unremarkable, magnitude 6.1 blue–white star. However, from a scientific perspective it is a superlative star, one of the most massive binary stars known, with a mass of between 40 and 110 times that of our Sun. Estimates do vary a bit in this regard, with the more modern numbers tending towards the upper end of the range. This being the case, Plaskett's Star is actually two very massive blue supergiants, one about 50 times the mass of the Sun, and the other about 40 times the mass of the Sun, each orbiting a common center of gravity with a period of a little over 14 days.

Perhaps the most astounding thing about Plaskett's Star is how close together the two components are – a mere 0.5 astronomical units separates them! To put that into perspective, Mercury is on average 0.39 astronomical units from the Sun, while Venus is around 0.72 astronomical units from the Sun. As bizarre as it sounds, these two giant stars are orbiting one another with a distance between them that is less than that of the distance between Venus and the Sun! Needless to say, these stars are too close together to be resolved at the eyepiece.

Because they are so close to one another, the speed at which they orbit one another is very fast. They are also close enough to be exchanging material; that is, the gravitational pull from each star is enough to pull away material from its companion. This extraordinary system won't be like this forever, but its precise fate remains a mystery. Big stars usually have dramatic endings, and that is surely the case here. Will one star throw out the other one eventually? Or will they both become neutron stars after they go supernova? Or maybe even a twinned pair of black holes? Nobody really knows. So although this star isn't much to look at through a telescope, it's an amazing star to ponder.

SAO 114258 (15 Monocerotis, S Monocerotis)

15 Monocerotis is an extraordinary multiple-star system with no fewer than sixteen components! Six components are visible through medium aperture telescopes. 15 Monocerotis A is a magnitude 4.7 star and has an intensely blue color. 15 Monocerotis B is a magnitude 7.5 star separated from 15 Monocerotis A by 2.8 arcseconds.

15 Monocerotis C, D, and E are much fainter stars, at magnitudes 9.8, 9.6, and 9.9, respectively. They are separated from 15 Monocerotis A by 16.6, 41.3, and 41.3 arcseconds respectively.

The sixth component, 15 Monocerotis F is brighter, a magnitude 7.7 star 2.6 arcminutes away from 15 Monocerotis A.

15 Monocerotis A is also an irregularly variable star, though not a dramatic one, it's brightness varying between 4.5 and 5.0.

SAO 117112 (Epsilon Hydrae)

This is a notoriously tricky multiple star system. There are actually five components in total, but it is Epsilon Hydrae A and Epsilon Hydrae B that are of interest. Epsilon Hydrae A is a yellow magnitude 3.4 star and Epsilon Hydrae C a blue magnitude 6.7 star. They are separated by slightly over 2 arcseconds, and when combined with the sharp difference in brightness, this closeness makes them difficult to split when viewed through small aperture telescopes.

SAO 131063 (Omicron-2 Eridani, 40 Eridani)

At magnitude 4.4, the orange star 40 Eridani A is the brightest star in a very interesting triple star system. It is a main sequence star a little smaller and cooler than our Sun, but like the Sun, it's a metal-rich star, implying that any planets that orbit it would likely be rich in the different chemicals required for life to evolve. The other two members of the system are 40 Eridani B, a magnitude 9.5 white dwarf and 40 Eridani C, a magnitude 11 red dwarf.

40 Eridani A is bright enough to be seen with the naked eye, but 40 Eridani B and C can only be seen with a telescope. Small telescopes will show the two companions as a single star; apertures upwards of 200 mm are needed to resolve them. 40 Eridani

B was the first white dwarf to be discovered and remains the easiest one in the sky for backyard astronomers to see.

This triple-star system is not just pleasant to look at; it's interesting to think about, too. In the *Star Trek* universe, 40 Eridani A is the star about which the planet Vulcan orbits, home of the redoubtable Mr. Spock. Although astronomers haven't yet found any Earth-like (terrestrial) planets orbiting 40 Eridani A, they have established that this star has a habitable zone within which such a planet might orbit. They've also established that 40 Eridani A is about 4 billion years old, a trifle younger than our own Sun, but still plenty of time for complex life, perhaps even intelligent life, to have evolved.

But is there life on 40 Eridani A? No one knows at the moment, but suffice it to say that 40 Eridani A is one of the ten targets for the proposed Terrestrial Planet Finder mission.

Since 40 Eridani A is part of a triple-star system, what might Vulcans see if they looked up at their skies? Would they see three suns instead of just one? No; 40 Eridani A would look like our Sun, but the other two stars would simply be very bright stars, brighter than Venus, but still point sources of light rather than discs. 40 Eridani B would, of course, be white, while 40 Eridani C would be red. Moreover, although these two stars would sometimes be visible during the day, they wouldn't provide appreciable amounts of heat or light, so they wouldn't be altering the weather or climate.

The constellation of Eridanus is relatively southerly, but even at far northern locations 40 Eridani shouldn't be too difficult to see. For observers in southern England it gets to about 30° above the horizon, high enough for it to be seen from anywhere that isn't too built up.

SAO 131907 (Beta Orionis, Rigel)

Rigel is a very bright (magnitude 0.2) supergiant star with 17 times the mass of the Sun and around 50,000 times its luminosity. Its sheer brilliance makes spotting its magnitude 6.8 companion rather tricky. Although the companion star is a respectable 9.5 arcseconds from the primary, Rigel is a tough split for medium aperture telescopes.

SAO 132314 (The Trapezium)

The Trapezium is a favorite wintertime target. Although often listed as a multiple-star system, Theta-1 Orionis C, the Trapezium is actually a small star cluster. The cluster contains at least eight stars, with four of them being bright enough to spot with small telescopes, and two more visible through medium to large aperture telescopes.

The four bright stars are components A, B, C, and D. These are white stars at magnitudes 6.7, 8.1, 5.4, and 6.7, respectively. Note that components A and B are both variable stars; A dims to magnitude 7.7 on a 65-day period, and B to magnitude 8.5 on a 6.5-day period. The four stars form a neat little quadrilateral shape, though more like a kite than a trapezium, with A at what would be top of the kite and D at the bottom.

Components E and F are both magnitude 10.0 stars and should be easily visible through medium aperture telescopes given favorable conditions. Component E is 4.5 arcseconds from A and F about the same distance from C.

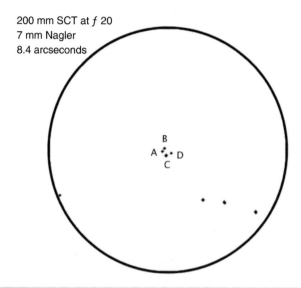

Figure 2.23. Four of the stars that make up the Trapezium are easily resolved; two others require medium to large apertures and high magnifications. (Image produced using Starry Night Pro. AllSky data courtesy of Main-Sequence Software Inc.)

SAO 132406 (Sigma Orionis)

Sigma Orionis is a famous five-star system with at least four stars that can be resolved with medium aperture telescopes. The brightest star (magnitude 3.7) is actually two very close stars, A and B, that can only be split with the very large aperture telescopes. The next brightest star is component E at magnitude 6.5 and 40 arcseconds from A/B. Component D is quite a bit fainter at magnitude 7.5, and 11.2 arcseconds away from A/B, nestling in between A/B and E. The faintest of the four resolvable stars is component C, a magnitude 10.3 star 11.3 arcseconds from A/B on the opposite side from D to E.

In the same field of view is the triple-star Struve 761, which you can slew to by locating SAO 132401 should you want to. The three stars in this system range in brightness from magnitude 7.9 to 8.7 and form a stretched triangle shape.

SAO 133317 (Beta Monocerotis)

Monoceras is a rather obscure constellation with only four stars above fourth magnitude and none above third magnitude, but the brightest of these stars, Beta Monocerotis, is a famous triple star. Indeed, some consider it to be the best triple star in the sky for amateurs using small and medium aperture telescopes.

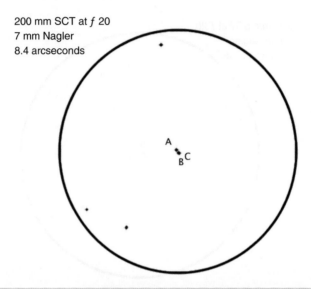

200 mm SCT at ƒ 20
7 mm Nagler
8.4 arcseconds

Figure 2.24. This obscure constellation contains a famous triple-star grouping. (Image produced using Starry Night Pro. AllSky data courtesy of Main-Sequence Software Inc.)

All three stars are essentially white in color, but one is noticeably brighter than the other two. The bright component is called Beta Monocerotis A, shines at magnitude 4.6, and is separated from the other two stars by about 7 arcseconds. Beta Monocerotis B and Beta Monocerotis C are separated by only 2.8 arcseconds and shine at magnitude 5.2 and 5.6, respectively. There is actually a fourth component almost half an arcminute away from Beta Monocerotis A, but it is too faint (magnitude 12.2) to be seen with the average amateur astronomer's telescope.

SAO 150058 (R Leporis, Hind's Crimson Star)

Hind's Crimson Star is, as its name suggests, remarkably red. Like many other blood-red stars it is a carbon star, a type of late stage giant star with a lot of carbon (as well as other heavy elements) in its atmosphere. In common with other carbon stars it is also a long-period variable star, changing in brightness from magnitude 5.5 down to 11.7 across a (somewhat irregular) period of roughly 430 days. Surprisingly, perhaps, most observers find that its reddish coloration is most apparent when it is at its dimmest.

SAO 150239 (Kappa Leporis)

Kappa Leporis consists of two blue–white stars, a magnitude 4.5 primary and a magnitude 7.4 secondary, the two stars separated by 2.6 arcseconds. This double star is considered a stiff challenge for small aperture telescopes, but should be easily split by medium to large aperture instruments.

SAO 151881 (Sirius, Alpha Canis Majoris)

Sirius is famous for being the brightest star in the sky (other than the Sun, of course). It is obviously blue–white today, but the influential Greek astronomer Ptolemy described it as being red, alongside other unambiguously red stars such as Antares and Betelgeuse. It's a bit of a mystery why he did so. One explanation is that he was viewing Sirius low down in the sky, and so its light was refracted by dust in the air and thereby appeared reddish. But that explanation isn't terribly satisfying. Ptolemy seems to have judged the colors of other stars very reliably. Since Ptolemy observed Sirius only 1,850 years ago, it doesn't seem likely that Sirius has substantially changed, either.

Sirius is a double star, but the small companion star, a white dwarf known as Sirius B, is so faint compared with its primary that it cannot be seen through most telescopes. Sirius B was deduced initially through the observations of the German astronomer Friedrich Bessel, who noticed that the proper motion of Sirius across the sky wasn't quite as it should have been. Essentially, it wasn't moving smoothly along a straight line, but wobbling ever so slightly. Bessel deduced that this was likely caused by the gravitational pull of an unseen companion. Bessel himself never saw the companion, and it would be almost 20 years before professional astronomers were able to observe it.

SAO 156110 (U Hydrae)

This is a strikingly red carbon star. It is a variable star as well, its brightness changing from 4.7 to 5.2 usually across 450 days but with some irregularity. This star has recently been photographed in the far infrared by the Japanese AKARI space telescope, and this revealed the presence of clouds of dust around the star. Seemingly this dying star is throwing out this material into interstellar space. The heavy elements in this stuff (i.e., the elements other than hydrogen and helium) will ultimately become the clouds of dust from which new stars are born, and things like carbon and nitrogen in particular will be crucial to the formation of complex chemicals and, perhaps, even life itself.

SAO 172676 (Epsilon Canis Majoris, Adhara)

At magnitude 1.5 this is the second brightest star in Canis Major. Although Sirius is much brighter from our vantage point on Earth, this is actually misleading. Of the two stars, Adhara is intrinsically by far the brighter. Astronomers describe intrinsic brightness by calculating how bright a given star would appear when viewed from a distance of 10 parsecs (about 32.6 light years). This is known as the absolute magnitude. For Sirius, this would be magnitude 1.4, and for Adhara, magnitude −4.1. Sirius seems so bright because it is very nearby, a mere 8.6 light years away, whereas Adhara is 430 light years away.

Stars move, of course, in what astronomers call their proper motion, but in most cases it takes thousands of years for changes in their relative positions to become obvious. But Adhara is interesting in this regard because it was once very much closer to Earth than it is today. If you were an astronomer viewing this star 5 million years ago, it would have shone at magnitude 3.99 because it was only 34 light years away. Beta Canis Majoris would also have been extremely bright and for the same reasons, only 37 million light years away and shining at magnitude −3.65. In both cases, these are comparable to the brightness of the planet Venus.

Adhara is of significance to professional astronomers as one of the strongest sources of ultraviolet in the sky. Amateur astronomers will be more interested in spotting its faint companion, a magnitude 7.5 star about 161 arcseconds from the primary. On paper the two stars sound like an easy split, but in practice it is much more difficult because the secondary is often overwhelmed by glare from the primary. Adhara is consequently a tough split for small to medium aperture telescopes, though some have split the two stars with as little as 75 mm telescopes.

SAO 198752 (Zeta Puppis)

Zeta Puppis is the hottest and bluest naked-eye star. It has a surface temperature of 40,000°C. At magnitude 2.2 it is a fairly bright star and easy to spot, but its southerly declination makes it difficult to see even at mid-northern locations. At far northern locations such as southern England it never rises above the horizon.

CHAPTER THREE

Spring

During the spring the orientation of Earth is such that when you look up at the night sky you are not looking into the center of the Milky Way galaxy but away from it. So instead of seeing those objects associated with the core of our galaxy, such as open clusters and globular clusters, you are instead looking at objects far outside our galaxy, in other words, other galaxies.

Astronomers in rural areas far from city lights often find this season highly rewarding, galaxies being amongst the most numerous and diverse deep sky objects. But for the backyard astronomer galaxies are very challenging targets. With a very few exceptions, galaxies are faint objects, and that means the astronomer wants both dark skies and a large-aperture telescope to see them well. Backyard astronomers will find that even low levels of light pollution can make galaxies very difficult to see, and things get even worse if you're using a small go-to telescope as well.

Unlike nebulae, light pollution filters aren't of any use because galaxies emit a broad range of light wavelengths rather than just a narrow band of them. So anything that cuts out light pollution is going to cut out light from the galaxy as well.

But it isn't completely hopeless! For a start, a properly aligned go-to telescope will at least place a galaxy in the field of view, removing the problem of finding the darn thing in the first place. Increasing magnification can also help, by enhancing the contrast between the galaxy and the background sky. While this also reduces the overall brightness of the image, meaning that fainter structures like the tips of spiral arms might vanish completely, you will at least be able to see the bright core of the galaxy. Depending on your telescope and the galaxy you're looking at, it's

N. Monks, *Go-To Telescopes Under Suburban Skies*, Patrick Moore's Practical
Astronomy Series, DOI 10.1007/978-1-4419-6851-7_3,
© Springer Science+Business Media, LLC 2010

worth trying magnifications between ×100 and ×200 before completely giving up on galaxy hunting.

Galaxies can be seen throughout the year, but the band of sky running from Ursa Major through Coma Berenices, Virgo and Leo is uncommonly rich in them. In fact this area of the sky contains what are known as galaxy clusters, groups of galaxies bound together by gravity. The Virgo Cluster is perhaps the best known, but there are many others, about which more is said in the Appendix.

Besides galaxies, spring is also the season when globular clusters start to make their appearance, being all but absent from the winter skies. Like galaxies, globular clusters look best under dark skies and through large-aperture telescopes. But globular clusters handle high magnifications very well, so upping the power to between ×150 and ×200 often works very well indeed. While the overall brightness will go down, the granular texture of globular clusters often becomes very apparent at high magnifications, and with averted vision you can resolve the better ones right down to the core, even under moderately light-polluted skies.

The Milky Way runs low along the horizon at this time of year, and your selection of nebulae and open star clusters is rather limited. Double stars feature a bit more strongly, so even if you don't have much luck bagging galaxies, you'll still find plenty of stuff to keep you amused in the springtime sky!

Showpiece Objects

NGC 3587 (M97, Owl Nebula)

One of the nicest planetary nebulae in the northern sky, the Owl Nebula is quite easy to spot even under suburban conditions. Small telescopes will probably show nothing more than slightly blurry 'star', but larger telescopes, around 200 mm in aperture, should show a bit more of its shape and structure. Under good conditions, you may even detect some greenish-blue color as well.

Light pollution filters dramatically improve the contrast, and are well worth using. Both the standard narrowband filter and the more selective O-III do a good job, though some observers maintain the O-III is ever so slightly better at teasing out detail.

The Owl Nebula gets its common name from two dark patches that resemble the eyes of an owl, the circular shape of the nebula itself suggesting the round face of the bird. Although obvious in photographs, seeing the two dark patches at the eyepiece is notoriously difficult. It's easy to fall into the trap of imagining you can see structural details like these just because you know they should be there. This delusion is probably a lot more common among amateur astronomers than they care to admit!

The Owl Nebula is about 2,600 light years away.

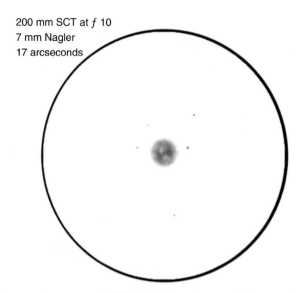

200 mm SCT at ƒ 10
7 mm Nagler
17 arcseconds

Figure 3.1. Use either a narrowband filter or an O-III filter to get the best views of the Owl Nebula. (Image produced using Starry Night Pro. AllSky data courtesy of Main-Sequence Software Inc.)

NGC 5194 (M51, the Whirlpool Galaxy)

The Whirlpool Galaxy in the constellation of Canes Venatici is a spiral galaxy about 23 million light years away. Visible through binoculars as a blurry star, and telescopes around 125 mm in size or larger should immediately reveal that the object actually contains not one but two bright spots: the core of the Whirlpool Galaxy itself, and the core of a smaller galaxy, NGC 5195, very close by. With a 200 mm or larger telescope, the spiral shape of the larger galaxy should be apparent, but this depends on the darkness of your skies. But even under suburban skies the bright core of the larger of the two galaxies should be easily visible.

In fact the two galaxies are interacting, NGC 5195 apparently passing through the edge of the Whirlpool Galaxy, producing the latter's distinct spiral shape in the process. Gravity also causes hydrogen molecules to clump together, and eventually some of these clouds of gas become so dense that they collapse under their own mass, becoming more and more compressed, and getting hotter and hotter, until nuclear reactions begin within their cores. In this way stars are born, and equivalent clouds of gas in our own galaxy include the Eagle Nebula M16 in Serpens that also happens to be visible at this time of year.

The Whirlpool Galaxy is the brightest galaxy within the cluster of galaxies named after its Messier designation, the M51 Galaxy Group. Most of the other members of this group are rather dim, but one exception is NGC 5055, the Sunflower Galaxy, also lying within the bounds of Canes Venatici.

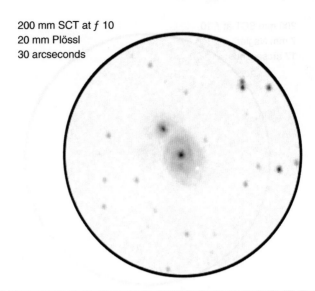

200 mm SCT at ƒ 10
20 mm Plössl
30 arcseconds

Figure 3.2. The Whirlpool Galaxy is a classic object for large telescopes under dark skies, but a challenging one for the suburban astronomer. (Image produced using Starry Night Pro. AllSky data courtesy of Main-Sequence Software Inc.)

NGC 5272 (M3)

This bright globular cluster is well positioned in the night sky during the second half of spring. Usually known to amateur astronomers by its Messier designation, M3, this globular cluster lies about 33,000 light years away from Earth and measures about 180 light years in diameter.

M3 it is to be found in the constellation of Canes Venatici. This region is notably devoid of bright stars, and amateur astronomers can find this globular cluster rather difficult to find using star hopping techniques. But as it lies halfway between the stars Arcturus and Cor Caroli, it can be spotted easily enough by sweeping the general area while using a wide-field, low-magnification eyepiece; M3 looks like moderately bright but fuzzy star.

Of course none of this really matters to owners of go-to telescopes! What does matter is that M3 is quite bright by globular cluster standards, and this lovely object can be relied upon to look good through any telescope. In terms of how condensed it is, M3 earns a middling Class VI rating on the Shapley–Sawyer scale, a nice balance between being compact enough to look good, but loose enough it's easy to resolve. Small telescopes will reveal its granular texture without much difficulty, but you will need an aperture of at least 114 mm to star resolving the outer layers of stars properly. Through a 200 mm telescope the view can be stunning!

If bagging dim galaxies has worn you out, M3 will be a first-rate pick-me-up. In fact the author considers M3 to be one of the best northern hemisphere globular clusters, and easily the equal of the better-known Great Hercules Cluster, M13.

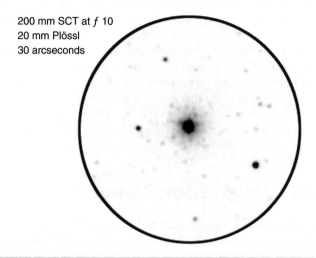

200 mm SCT at ƒ 10
20 mm Plössl
30 arcseconds

Figure 3.3. M3 is one of the best globular clusters in the northern sky, and a splendid sight through a 200 mm SCT. (Image produced using Starry Night Pro. AllSky data courtesy of Main-Sequence Software Inc.)

NGC 5904 (M5)

M5 is another lovely globular cluster, albeit one that isn't well placed until the end of spring. It is about 160 light years in diameter and lies some 25,000 light years from Earth. While broadly similar in these particulars to the globular cluster M3, M5 seems to contain much older stars. Current estimates place its age at around 13 billion years, as opposed to a mere 8 billion years for the globular cluster M3.

M5 is easily bright enough to see with binoculars. On the Shapley–Sawyer scale it earns a middling Class V rating, and all but the smallest telescopes will reveal its granularity. Though perhaps (very slightly) inferior to M3 in terms of visual oomph, M5 is still a superb globular cluster that stands up well to high magnifications.

NGC 6543 (Cat's Eye Nebula)

Located in the dim, northerly constellation of Draco, is the Cat's Eye Nebula, a planetary nebula about 3,300 light years away. It is small but comparatively bright, and generally easy to find, though at first glance it is easily mistaken for a slightly out of focus star.

Through most telescopes the Cat's Eye Nebula looks like a somewhat blurry blue–green star about 20 arcseconds in diameter. Even a 200 mm aperture will reveal little detail, and you'll need a telescope about twice as big to see the dying star at the center of the nebula. Long-exposure photographs reveal the nebula to be an amazingly complex structure consisting of helical whorls of hot gas circling a magnitude 11 star.

Both narrowband and O-III filters work wonders on this object, making it easy to spot even in mildly light-polluted skies. If you're not sure which 'star' is the Cat's Eye Nebula, try putting a light pollution filter between your eye and the eyepiece: the Cat's Eye Nebula is the object that doesn't become noticeably dimmer.

One piece of astronomical trivia is that the Cat's Eye Nebula lies almost exactly at the North Ecliptic Pole. If you visualize the night sky as a globe, with the ecliptic as the equator, the Cat's Eye Nebula would be at the north pole.

NGC 6611 (M16, Eagle Nebula)

Towards the end of spring, the constellation of Serpens becomes easier to explore, and one of its many treasures is the deep sky object known as the Eagle Nebula. It is a complex object about 7,000 light years away that contains an open star cluster surrounded by a faint emission nebula.

The star cluster part of this object is bright and moderately large, and generally easy to see. But the nebulosity is very difficult to spot, and you'll need a large aperture

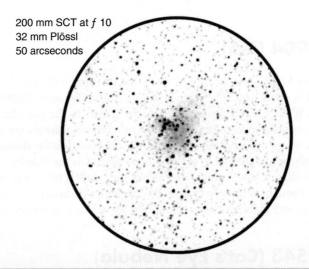

200 mm SCT at ƒ 10
32 mm Plössl
50 arcseconds

Figure 3.4. The L-shaped star cluster within the Eagle Nebula is easy to see under suburban sky conditions, but the nebula part of the object is tricky. Narrowband and O-III filters help significantly. (Image produced using Starry Night Pro. AllSky data courtesy of Main-Sequence Software Inc.)

and dark skies to see much detail. Narrowband and O-III filters will be of great use to the suburban observer. While they won't replicate the dark sky experience under light polluted skies, they will at least make the difference between spotting the nebula and not seeing anything there at all.

In recent years photographs of the Eagle Nebula taken by the Hubble Space Telescope have become well known, in particular the photograph known as the 'Pillars of Creation'. Despite that Biblical-sounding name, the Eagle Nebula is a relatively young object, and an active site of star formation.

Interesting Deep Sky Objects

NGC 2841 (Tiger's Eye Galaxy)

The Tiger's Eye Galaxy is a frequently overlooked but rewarding spiral galaxy in the constellation of Ursa Major. Although this galaxy is about 46 million light years away from Earth, it's quite bright and easy to see. Provided the light pollution isn't too severe, it should be visible as a somewhat elongated oval blur of light with a noticeably bright, almost star-like, core. Under good conditions dust lanes can be observed, but these likely won't be visible from the average suburban location. Seeing its spiral arms is even more of a challenge.

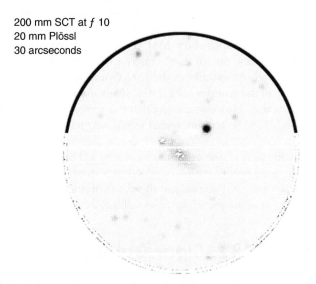

200 mm SCT at *f* 10
20 mm Plössl
30 arcseconds

Figure 3.5. The bright core of the Tiger's Eye Galaxy should be visible under moderately light polluted conditions, its spiral shape and prominent dust lanes will be rather more difficult to see. (Image produced using Starry Night Pro. AllSky data courtesy of Main-Sequence Software Inc.)

Compared to our own Milky Way galaxy, the Tiger's Eye Galaxy is about 50% larger, spanning a diameter of 150,000 light years rather than our own galaxy's 100,000 light years. Over the years astronomers have observed a number of super-novas in this galaxy.

NGC 2903

This moderately bright galaxy in Leo is about as good as northern hemisphere non-Messier Catalogue galaxies get. It is a spiral galaxy, but we're viewing it at an angle, so this is less obvious than with those spiral galaxies that we view face on. Still, under reasonably dark skies and with an aperture of at least 200 mm, you should be able to detect the arms as a certain haziness around the bright core.

NGC 2903 is a spiral galaxy about 20 million light years away. Curiously, a particularly bright cloud of stars within the galaxy has its own NGC designation, NGC 2905. This feature is more obvious in photographs than at the telescope eyepiece.

NGC 3031 (M81, Bode's Galaxy)

Sometimes known as Bode's Galaxy, this bright galaxy is in the constellation of Ursa Major and lies about 11 million light years way. It is spiral galaxy with a large core, clearly elliptical rather than circular. Its spiral arms are notoriously difficult to observe to observe though. Most observers find them difficult to detect even under dark skies when using telescopes of aperture 200 mm or less.

Long-exposure photographs show M81 to be a stunning example of a spiral galaxy, and it has become one of most closely studied galaxies in the heavens. It is one of the two brightest galaxies in the M81 Group and by far the biggest. The other bright galaxy in the group is M82 (NGC 3034) and in fact the two galaxies are known to be interacting. M81 is so massive that it is severely distorting M82, causing star formation in M82 to be ten times what would normally be expected in a galaxy of its type.

Low power, wide field eyepieces should be able to put both galaxies into a single field of view. Aim for a field of view around a degree in width for best results. A 32-mm Plössl on a 200-mm SCT should just squeeze them in, but if not, use a reducer-corrector as well to decrease the telescope's focal length.

NGC 3034 (M82, Cigar Galaxy)

M82 is a believed to have originally been a spiral galaxy, but it has been so distorted by the nearby M81 galaxy that little of its original shape remains. It is now classified as an irregular galaxy.

About 100 million years ago M82 entered the gravitational field of its massive neighbor, and gravitational forces triggered ten times the amount of star formation that would be typical for this class of galaxy. Distorted galaxies exhibiting dramatically higher levels of star formation are known to astronomers as starburst galaxies, and M82 is one of the easiest examples of the amateur astronomer to observe.

The sites of star formation are not visible through small telescopes, and in any case, most of the evidence for star formation comes from telescopes sensitive to x-rays, infrared light, and radio frequencies. But the amateur should still be able to see that M82 is a lumpy-looking galaxy, and its crooked shape stands in sharp contrast to the nicely elliptical M81. Telescopes with an aperture of 200 mm or more should reveal the dark lane of dust running along the middle of this galaxy, but this isn't an easy feature to see.

Like M81, M82 is about 11 million light years away.

NGC 3115 (Sextans Spindle Galaxy)

The Spindle Galaxy is a bright galaxy in Sextans that tends to be overlooked because of its location so far south of Leo. Observers at mid to high northern latitudes – such as those in Canada or northern Europe – may find that this galaxy barely rises above the southern horizon, if it does so at all. But observers further south should get much better views, as will observers in the southern hemisphere.

Assuming you're lucky enough to be able to see the Spindle Galaxy, you'll be rewarded with an edge-on view of a particularly nice lenticular galaxy. As its name suggests, it has a bulging center that quickly tapers off into long, thin extensions on either side. To be fair, this won't be obvious to observers under light-polluted skies or those using small-aperture telescopes, but even these observers should be able to spot the bright central core.

The Spindle Galaxy is about 32 million light years away from us. It is a rather odd galaxy in some ways, apparently having a dormant black hole in the center and exhibiting little sign of recent star formation.

A quick word of warning: this isn't the only Spindle Galaxy in the sky! The name has also been used to describe NGC 5866, another lenticular galaxy, this time in the constellation of Hydra (and discussed elsewhere in this book).

NGC 3351 (M95)

This reasonably bright galaxy should be easy to spot under dark skies, but can be a challenge for suburban astronomers. It is generally considered to be among the more difficult objects on the Messier list, and realistically, suburban astronomers are going to need at least 114 mm for a decent view of this galaxy.

Upping the magnification can help, but the downside to that approach is you lose one of the best things about this galaxy, it's close proximity to M96 (NGC 3368). The two galaxies lie about 40 arcminutes apart, so if you want to try see both objects, use the lowest magnification you can. With a 200-mm SCT, a 32-mm Plössl will just

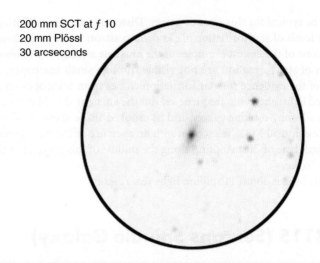

200 mm SCT at *f* 10
20 mm Plössl
30 arcseconds

Figure 3.6. The Sextans Spindle Galaxy is fairly bright and the core at least can be seen readily through moderate light pollution. (Image produced using Starry Night Pro. AllSky data courtesy of Main-Sequence Software Inc.)

about do, but if you can use a reducer-corrector to decrease the focal length as well, so much the better.

M95 is a barred spiral galaxy that lies about 33 million light years away, and is a member of a group of galaxies known as the M96 or Leo I Group.

NGC 3368 (M96)

The other relatively bright member of the M96 Group of galaxies is of course M96 itself (NGC 3368). Some observers maintain that it is marginally easier to see than M95, but there isn't much in it, both being challenging objects for those under suburban skies. Large telescopes and dark skies reveal a certain amount of mottling, but suburban astronomers will probably have to be satisfied with simply spotting this object, let alone discerning any features or structure.

M96 lies about 31 million light years away. It isn't quite a barred spiral galaxy in shape but it isn't quite an unbarred spiral galaxy either, and is therefore called an intermediate spiral galaxy.

NGC 3623 (M65)

M65, M66, and NGC 3628 make up a small group of galaxies known as the Leo Triplet, lying about 35 million light years away. There's some debate over whether the Triplet are their own thing or merely part of a much larger group, with some astronomers maintaining that they are outlying members of the M96 Group.

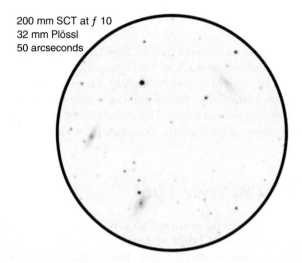

200 mm SCT at ƒ 10
32 mm Plössl
50 arcseconds

Figure 3.7. With a suitably low power, wide field eyepiece, all of the Leo Triplet of galaxies can be seen in the same field if view. M65 (*at left*) and M66 (*bottom left*) are fairly bright and easy to spot; NGC 3628 (*top right*) is fainter and more difficult to see. (Image produced using Starry Night Pro. AllSky data courtesy of Main-Sequence Software Inc.)

Be that as it may, the Leo Triplet is a popular target for amateur astronomers, the two brighter members, M65 and M66, being easily spotted even in relatively small telescopes. They are only separated by about 20 arcminutes, and a 200-mm SCT equipped with a 32-mm Plössl should comfortably place them in the same field of view. NGC 3628 is about 35 arcminutes from both M65 and M66, and is rather faint, but under reasonably dark skies you should be able to squeeze it into the field of view as well.

M65 is an intermediate spiral galaxy that appears as a slightly dimmer, more oval blur than M66.

NGC 3627 (M66)

M66 is a member of the Leo Triplet, and can be easily distinguished from M65 by its slightly higher brightness and more elongated shape. Like M65, it's an intermediate spiral galaxy. Under good conditions, it should be apparent that M66 isn't uniform in brightness, with a central bar-like region of higher brightness running through the middle of the galaxy.

M66 is only 20 arcseconds away from M65, and a low power eyepiece should place them both into the same field of view. M66 is difficult to see in telescopes less than 100 mm in aperture under suburban skies, but shouldn't be too difficult for telescopes in 150–200 mm aperture range.

NGC 3628

The faintest of the three members of the Leo Triplet, this galaxy requires fairly dark skies as well as reasonably generous aperture to be easily observed.

At some point in time, M66 and NGC 3628 interacted with each other, and among other things, part of NGC 3628 was drawn out into a long 'tail' measuring a whopping 300,000 light years in length! This so-called tidal tail of gas, dust and stars is visible on long-exposure photographs, but cannot be seen through telescopes.

NGC 4258 (Messier 106)

This fairly bright galaxy lies in the constellation of Canes Venatici about 24 million light years from Earth. It is the brightest members of a group of galaxies known as the M106 Group or Canes Venatici II Group. Most of the other members are much dimmer than M106, and consequently the group isn't much known among amateur astronomers.

M106 is a spiral galaxy of a type known as a Seyfert galaxy. Such galaxies have very bright cores for their size, emitting not just visible light but also radio waves, infrared light, ultraviolet light and x-rays. While the details are still unclear, it's believed an accretion disk around a central black hole is responsible for this brightness.

You can't see any of this through a telescope of course, but an aperture of at least 100 mm should reveal its basic shape, that of an elongate blur with a relatively bright core. Its arms are very difficult to see under suburban skies, even with a 200 mm aperture telescope.

NGC 4361

This planetary nebula is one of very few deep sky objects in the constellation of Corvus easily visible through amateur telescopes. It's reasonably bright and not too difficult to spot, though under low magnification it may appear to be nothing other than a fuzzy star. Large apertures and higher magnifications should reveal a little more detail, and some observers have likened the shape of this nebula to that of a spiral galaxy.

Under dark skies the central star is visible through apertures down to 100 mm; suburban observers will need much more aperture than this. Because it has used up virtually all of the fuel that drives its nuclear reactions, this star is cooling down and shrinking; it is becoming a white dwarf. Such stars don't evolve any further, just becoming cooler and cooler as they age, emitting less and less light. In theory a white dwarf would eventually reach the same temperature as the interstellar medium, at which point they'd not emit any light at all, but so far as we know, the universe isn't old enough for any white dwarf to have cooled down that far.

A light pollution filter is helpful when observing NGC 4361, increasing its contrast while dimming the stars around it. There's not much difference between how a narrowband filter performs versus an O-III filter, so use whichever one you have handy.

NGC 4472 (M49)

M49 is an elliptical galaxy that is relatively easy to spot under reasonably dark suburban skies. Through the average amateur's telescope little structure will be seen, but long-exposure photographs have revealed a number of small companion galaxies around its edges, as well as a system of globular clusters. One of these globular clusters is home to a small black hole, one that weighs about ten times the mass of our Sun. Astronomers believed that globular clusters would tend to kick out, rather than retain, black holes, so the discovery of this particular black hole was rather a surprise. So while NGC 4472 isn't the most rewarding galaxy to look at through a telescope, it's an interesting galaxy nonetheless.

M49 is located about 60 million light years away from Earth and the brightest member of the Virgo Cluster of galaxies. The Virgo Cluster has been divided into three separate regions called subclumps, and M49 is at the center of one of them, the other two subclumps being centered around M87 and M86. There may be other subclumps as well, but these are three largest. Each subclump consists of galaxies gravitationally bound to one another, and the three subclumps in turn gravitationally bound to each other.

NGC 4486 (M87)

M87 is a giant elliptical galaxy in the constellation of Virgo. It measures some 120,000 light years in diameter, making it quite a bit bigger than our own Milky Way Galaxy, which has a diameter of about 100,000 light years. Another interesting contrast may be drawn by counting its collection of globular clusters. M87 has something like 12,000 of them, whereas the Milky Way has less than 200.

This is one of the galaxies for which there is good evidence for the presence of a supermassive black hole at its center. In the case of M87, the most striking evidence is a stream of matter more than 5,000 light years in length has been detected flying out of the core. M87 is also a very strong emitter of radio waves, x-rays, and gamma rays, and these may also be related to presence of a black hole.

Needless to say, amateur astronomers won't see any of this, but they will be rewarded with a view of a bright, roughly circular galaxy easily viewed under all but the worst suburban skies.

M87 is one of the largest members of the Virgo Cluster and is located some 55 million light years away.

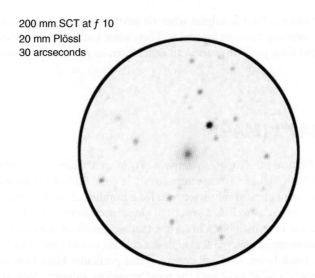

200 mm SCT at ƒ 10
20 mm Plössl
30 arcseconds

Figure 3.8. M87 is a giant elliptical galaxy that is reasonably easily seen under suburban skies. (Image produced using Starry Night Pro. AllSky data courtesy of Main-Sequence Software Inc.)

NGC 4490 (Cocoon Galaxy)

NGC 4490 is the brighter of two interacting galaxies, the other being NGC 4485. They lie about 45 million light years away, and are moving away from each other after. A thin, wispy tail of stars connects the two galaxies, and this region appears to be an area of intensive star formation.

Technically they're spiral galaxies, but even in photographs that isn't obvious because of how their gravitational fields have distorted one another. Although fascinating objects, they are notoriously difficult to see under suburban skies. Simply spotting NGC 4490 is difficult, let alone its much smaller and fainter companion. The bright core of NGC 4490 should be visible even through telescopes around the 100 mm aperture mark, but to spot NGC 4485 requires considerably more aperture.

NGC 4494

Although not a particularly impressive galaxy, the location of this elliptical galaxy in the Coma Berenices Cluster (also known as Melotte 111) is pretty neat. It's only half a degree from the 17 Comae Berenices, a double star with a white primary and a blue companion.

NGC 4494 is about 32 million light years away. It looks like a small, faint globular cluster, and even fairly large aperture telescopes don't reveal much detail.

NGC 4565

One of the most attractive galaxies in Coma Berenices, this spiral galaxy is viewed edge-on, and its bright core is fairly easy to spot. With sufficient aperture its proper shape should become apparent, as the disc reveals itself as narrow, almost needle-like extensions on either side of the core. Long-exposure photographs clearly show this galaxy's prominent dust lanes, and observers with large aperture telescopes can glimpse them under dark skies. These dust lanes are, unfortunately, very difficult to see under suburban conditions.

NGC 4565 is about 20 million light years away and a member of the Coma Cluster group of galaxies.

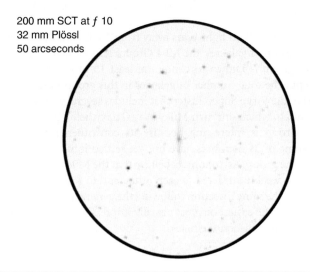

200 mm SCT at ƒ 10
32 mm Plössl
50 arcseconds

Figure 3.9. Because NGC 4565 is viewed edge-on, its surface brightness is comparatively high, and consequently its shape is relatively easy to see. (Image produced using Starry Night Pro. AllSky data courtesy of Main-Sequence Software Inc.)

NGC 4579 (M58)

M58 is a barred spiral galaxy in the constellation of Virgo. It is quite bright and easy to spot, though it reveals little detail, even at high magnifications and through large aperture telescopes. A 200-mm SCT should at least show a hint of its spiral arms, but that's about all you can hope for under suburban skies.

This galaxy is a member of the Virgo Cluster and located about 68 million light years away. At least two supernovae have been observed in this galaxy.

NGC 4736 (M94)

Although this spiral galaxy in Canes Venatici is fairly bright, it reveals little of its shape through small to medium aperture telescopes. At best all that can be seen under suburban conditions or through a small telescope is that the bright core in the middle is surrounded by a more or less uniformly faint halo.

Long-exposure photographs reveal delicate spiral arms, as well as two distinctive ring-like structures. One is relatively close to the core, the other much further out. These rings are very difficult to see visually, and to do so will require very dark skies and a large-aperture telescope. These rings are occasionally seen on other galaxies, too, such as the southern hemisphere galaxy NGC 1512 in Dorado, but M94 is the one where they are best known and most studied. The rings appear to be sites of intense star formation, but what forms the rings in the first place remains unclear.

M94 is about 16 million light years away from Earth and the eponymous member of its own group of galaxies, the M94 Group, also known as the Canes Venatici I Group. This group is known to contain at least 15 other galaxies, though some astronomers put the total number of galaxies in this group as high as 24. The M94 Group is part of the Virgo Supercluster that includes several galaxy groups, including the Local Group to which our own Milky Way Galaxy belongs.

The M94 Group is interesting because its constituents are relatively loosely bound, and some of its members have not yet settled into orbits around the common center of the group. Astronomers believe that the M94 Group contains galaxies that are still only weakly attracted to each other, and to some extent merely happen to be traveling in the same direction following the general expansion of the universe. So while M94 isn't the easiest or most visually impressive deep sky object in the sky, it's a very interesting object nonetheless.

NGC 4826 (M64, Black Eye Galaxy)

The Black Eye Galaxy is one of the galaxies in Coma Berenices most appreciated by amateur astronomers. But although it's a great target for observers under dark skies and large-aperture telescopes, it can be a disappointment to suburban observers, particularly those with relatively small telescopes.

This galaxy gets its common name from its distinctive appearance when viewed under good conditions. Like other spiral galaxies it has a bright central core surrounded by a dimmer halo, but along one side of the core the halo is obscured by a dark curved band. The result is something a bit like a human eye with a bruise underneath it, hence Black Eye Galaxy. Of course, this isn't a bruise, and the dark band we see, in fact, consists of at least two clouds of dust that are blocking off the light being emitted by the stars behind them.

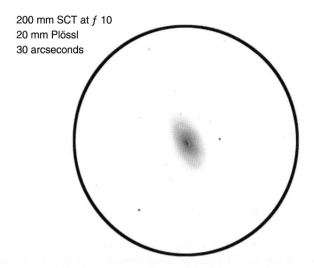

200 mm SCT at ƒ 10
20 mm Plössl
30 arcseconds

Figure 3.10. Despite its fame among amateur astronomers, the Black Eye Galaxy is not easy to see from suburban locations. Its bright core should be visible, but spotting the dust lanes that provide its popular name usually requires quite dark skies. (Image produced using Starry Night Pro. AllSky data courtesy of Main-Sequence Software Inc.)

Even under suburban conditions the bright core should be apparent, even if little else can be seen. But sadly it really does take dark skies and relatively high magnifications for the black eye part of the galaxy to become apparent, and as is often the case with discerning galaxy structures, you have to be careful not to imagine you can see details you know to be there.

Although you might not be able to see much detail, you can at least appreciate this galaxy for being remarkable in several ways. But whereas most galaxies contain stars all orbiting the center in the same direction, the inner region of the Black Eye Galaxy rotates one way, and the outer region the other! Where the two regions meet, about 3,000 light years out from the center of the galaxy, there are massive amounts of star formation as clouds of gas collide and become compressed.

Astronomers believe that about a billion year ago the Black Eye Galaxy collided with, and absorbed, another galaxy. The result was that while the central part of the Black Eye Galaxy continued to rotate it the normal way, its outer region, now largely made up of stars from the other galaxy, ended up rotating in the other direction, the wrong direction by comparison with most other galaxies. All in all, a bizarre galaxy, well worth adding to any observing list for this time of the year.

There's some debate about how far away the Black Eye Galaxy is, with estimates ranging from 10 to 40 million light years, though a figure of around 20 million light years is now widely accepted. The Black Eye Galaxy is a member of the M94 Group, also known as the Canes Venatici I Group.

NGC 5024 (M53)

Located in Coma Berenices, a constellation best known for its galaxies, M53 is one of the more distant globular clusters at a distance of around 60,000 light years from the center of the Milky Way Galaxy. It is relatively small and faint, but there aren't many globular clusters visible at this time of the year, so it's well worth spending a little time with this object, playing around with magnification to see if you can tease out a little more detail. Under good conditions it's an attractive object, though fairly condensed, Class V on the Shapley–Sawyer scale, and you'll need a telescope upwards of 200 mm to resolve it down to the core.

NGC 5055 (M63, Sunflower Galaxy)

This celebrated galaxy in Canes Venatici was one of the first for which a spiral shape was discerned. It was one of a group of such objects identified by William Parsons, the Third Earl of Rosse during the late 1840s from his observatory in Ireland. At the time such objects were known as spiral nebulae, but we know them now as spiral galaxies.

The spiral shape is best seen in photographs, but you can see them easily enough under dark skies through telescopes with an aperture of 150 mm or more. Its spiral arms aren't neatly arranged though, but appear lumpy and haphazardly arranged. Its common name refers to this, the suggestion being that the spiral arms are like the clustered petals around the center of a flower. This is all largely academic for suburban astronomers though, who won't be able to see the spiral arms at all. The central core is pretty bright though, and should be easy enough to spot.

The Sunflower Galaxy is a member of the M51 Galaxy Group and lies about 37 million light years away from us.

NGC 5457 (M101, Pinwheel Galaxy)

This lovely galaxy in Ursa Major is known as the Pinwheel Galaxy because it appears face-on to us, and we can see the central core and the orbiting spiral arms very clearly, should conditions allow. And that's the tricky bit: should conditions allow. Although this is a fairly bright galaxy, its face-on orientation means that the light it emits is spread out across a wider area than an edge-on galaxy, so it isn't all that easily spotted against the background sky. The bright core should be visible on dark nights through all but the smallest instruments, but seeing the spiral arms is very difficult except under the very best skies. Usually all that can be seen is a uniform faint blur around the bright core.

The Pinwheel Galaxy is often given as an example of a Grand Design spiral galaxy, that is, a galaxy with a strong division between the bulging core and flat, clearly-defined spiral arms. Although that's a fair description of the galaxy in general terms, its arms are not mirror images of each other, and in fact one arm is substantially thicker and brighter (more star-rich) than the others. Astronomers think that at some point in the distant past the Pinwheel Galaxy passed close by another galaxy, and while the two didn't collide, gravitational forces were sufficient to disrupt the spiral arms, resulting in their uneven appearance.

The Pinwheel Galaxy is about 27 million light years away. It is the brightest member of a group of galaxies known as the M101 Group, which in turn is a member of the Virgo Supercluster (also known as the Local Supercluster, since it includes the Local Group that contains our own Milky Way Galaxy).

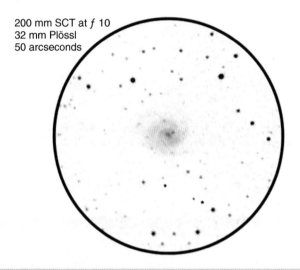

200 mm SCT at ƒ 10
32 mm Plössl
50 arcseconds

Figure 3.11. Like a lot of spiral galaxies viewed face-on, M101 is a faint and difficult object for observation under suburban skies. That said, the bright core should be easily visible. (Image produced using Starry Night Pro. AllSky data courtesy of Main-Sequence Software Inc.)

NGC 5466

This globular cluster in Boötes is almost 52,000 light years from Earth and one of the more difficult globular clusters of this season. It is quite large, but because of its size, it is also rather dim, so seeing it against a light-polluted sky isn't easy. Indeed, under suburban conditions it may well be completely invisible.

NGC 5466 is one of the loosest globular clusters in the sky. The Shapley–Sawyer classification scale runs from I to XII, I being the most condensed, and XII being the least condensed. NGC 5466 falls comfortably into the XII category, and at the eyepiece it can look more like a dim open cluster than a globular cluster. Given its size and lack of brightness though, this is a disappointing object for astronomers will small aperture telescopes.

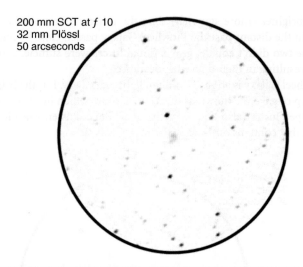

200 mm SCT at ƒ 10
32 mm Plössl
50 arcseconds

Figure 3.12. NGC 5466 is a large but dim globular cluster. (Image produced using Starry Night Pro. AllSky data courtesy of Main-Sequence Software Inc.)

Obscure and Challenging Deep Sky Objects

NGC 3079

One of the less easily observed galaxies in Ursa Major, NGC 3079 is a moderately bright galaxy oriented edge-on to us, so all that can be seen at the eyepiece is thin, somewhat irregular patch of light. The core is somewhat brighter than the arms, and should appear as a thicker, brighter bulge in the center of the patch.

NGC 3079 is sometimes called the Bubbling Cauldron Galaxy because of a peculiar feature visible in long-exposure photographs. Towers of hot gases some 2,000 light years in height are streaming away from the core, which astronomers have likened to the bubbles and steam rising above a boiling cauldron. Why these structures are present isn't understood, though one idea is that streams of particles are being produced by massive stars in the core of the galaxy. These streams of particles would be somewhat like the solar wind that blows through our Solar System, but on a far larger scale. But for this explanation to make sense, the center of NGC 3079 would have to contain an area of star formation producing a lot of very big stars.

NGC 3184

This is one of several galaxies in Ursa Major visible through moderate aperture telescopes. However, because NGC 3184 is viewed face-on, the light it emits is comparatively spread out, and even in quite big telescopes it's rather a dim object. Long-exposure photographs reveal this galaxy to be an nice example of the spiral galaxy type, but visual observers will usually only see a dim, roughly circular blur.

NGC 3184 is about 25 million light years away and is a member of a small group of galaxies that also contains NGC 3104, NGC 3198, and NGC 3319. A small number of supernovas have been observed in this galaxy, including one in 1999.

NGC 3379 (M105)

M105 is an elliptical galaxy in Leo that is bright enough to see even with relatively small telescopes under dark skies. At low magnifications, the elliptical nature of this galaxy isn't obvious, and it appears to be a blurry but fairly bright star. Higher powers will reveal a somewhat circular shape, almost like a globular cluster. Long-exposure photographs reveal a dust lane that runs across the center of the galaxy slightly to one side of the core. Astronomers have also learned that this galaxy is one that contains a supermassive black hole at its center.

M105 is popular with amateur astronomers because another galaxy, NGC 3384, can be caught in the same field of view. The two galaxies are only about 7 arcminutes apart, and if M105 is in the center of the field of view, NGC 3384 shouldn't be too far away. If a 20-mm Plössl is used on a 200-mm SCT for example, NGC 3384 will be about halfway between M105 and the edge of the field of view. But be warned: NGC 3384 is a little bit fainter than M105 and easily overlooked.

Another, even dimmer galaxy, NGC 3389, should also be in the field of view, the three galaxies forming the points of a roughly equilateral triangle. It is much dimmer than the other two galaxies, and may completely invisible under suburban skies.

M105 and NGC 3384 are both members of the same group of galaxies, the M96 Group of galaxies, sometimes known as the Leo I Group. M105 is about 32 million light years away. By contrast, NGC 3389 isn't part of this group at all, and lies at least twice as far away. It simply happens to be in the background of this particular field of view.

NGC 3384

NGC 3384 is a lenticular galaxy in Leo. It appears to be a very ancient galaxy, its constituent stars being overwhelmingly of the metal-poor Population II stars typically found on globular clusters.

This galaxy is slightly fainter than its close neighbor M105, and the two galaxies are fun to contrast when viewed at the same time. A wide-angle or low-magnification eyepiece should do the trick, the two objects being separated by only about 7 arcminutes.

Besides being fainter, NGC 3384 also has a different shape and appears distinctly elongated compared with the nearly circular M105. Both galaxies are members of the M96 Group, though NGC 3384 is a little further away than M105, at about 35 million light years distant.

NGC 3521

This bright spiral galaxy in Leo is visible with small aperture telescopes under good conditions, and should be easily visible to those using 150–200 mm apertures under suburban skies. Its core is much brighter than its arms though, and unless your skies are reasonably dark, the core may be only thing readily visible. Really big telescopes (250 mm upwards) under dark skies reveal quite a bit more detail, including spiral arms and a certain degree of mottling.

NGC 3521 is about 32 million light years away.

NGC 3556 (M108)

This bright galaxy in Ursa Major is a peculiarity: a spiral galaxy without a swollen central core. From Earth it is viewed edge-on, and yet it looks more or less like a somewhat irregular elongate blur. Under good conditions the galaxy can be seen to be mottled, evidence of the substantial dust lanes that this galaxy possesses.

M108 is about 45 million light years away and is one of 32 large galaxies, and numerous smaller ones, that make up the Ursa Major North Group, also known as the M109 Group.

NGC 3607

NGC 3607 is a lenticular galaxy in Leo about 88 million light years away. It is quite bright and easy to see with medium to large aperture telescopes. Not much detail is visible in this galaxy, but it's worth tracking down because the nearby galaxy NGC 3608 should be visible in the same field of view, the two galaxies being separated by only 5 arcminutes.

This galaxy is the brightest member of the 110 galaxies that make up the Leo II Groups, sometimes called the Leo II Cloud. Whereas the Leo I Group is a single lump of galaxies all close together, the Leo II Groups are spread out across a vast tract of space, and consequently astronomers treat them as multiple smaller groups rather than a single big one. Taken together, the Leo II Groups are a major component of the Virgo Supercluster.

NGC 3877

Although quite bright, there's a trick to seeing this galaxy. Because it's so near the star Chi Ursae Majoris, a mere 15 arcminutes away in fact, if you leave this magnitude 3.7 star in the field of view, the galaxy will be virtually impossible to see. Since your go-to telescope will have centered the galaxy automatically, try changing the magnification by swapping eyepieces if Chi Ursae Majoris is in the field of view. For a 200-mm SCT, a 20-mm Plössl eyepiece should center the galaxy comfortably while excluding Chi Ursae Majoris.

This galaxy is about 50 million light years away and a member of the Ursa Major North Group of galaxies, also known as the M109 Group.

NGC 3941

This galaxy is of the barred lenticular type. Like barred spiral galaxies, barred lenticular galaxies have a central bar-like structure that crosses the core, but lenticular galaxies have much less obvious spiral arms than spiral galaxies, which is why they form their own class in between spiral galaxies on the one hand and elliptical galaxies on the other.

NGC 3941 is further distinguished by being one of the Seyfert-type galaxies, meaning that they have very bright cores and emit distinct wavelengths of light. They are believed to contain supermassive black holes. In the case of NGC 3941, its peculiarities may be because this galaxy is actually what was produced when two separate galaxies collided and combined.

NGC 3941 lies about 50 million light years away.

NGC 3992 (M109)

M109 is one of the more difficult Messier galaxies to see. It is a barred spiral galaxy some 55 million light years away, and it takes good skies and a reasonably large aperture telescope (upwards of 200 mm) for the central bar to be visible. Under less promising conditions or with smaller telescopes the bright core at least should be apparent, the somewhat hazy and elongated disc being a bit more difficult to spot.

M109 belongs to the Ursa Major North Group of galaxies, sometimes known as the M109 Group after this, its brightest member.

NGC 4026

NGC 4026 is a lenticular galaxy in Ursa Major some 60 million light years away. Under good conditions it should be visible as a thin, spindle-shaped patch of light with a much brighter core. Although not a particularly dramatic object, it is at least reasonably easy to see under suburban conditions.

This galaxy is a member of the Ursa Major North Group of galaxies, also known as the M109 Group.

NGC 4038 and NGC 4039 (Antennae Galaxies)

These two barred spiral galaxies in the constellation of Corvus form a striking pair of objects. Assuming reasonably dark skies, medium to large aperture telescopes should reveal a V-shaped object of approximately uniform brightness.

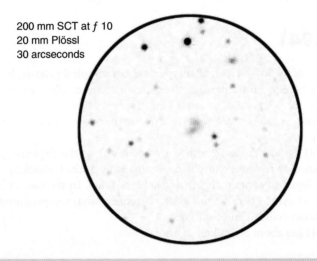

200 mm SCT at *f* 10
20 mm Plössl
30 arcseconds

Figure 3.13. The Antennae Galaxies are difficult to observe under light-polluted conditions, though their cores may be apparent as faint V-shaped blur. (Image produced using Starry Night Pro. AllSky data courtesy of Main-Sequence Software Inc.)

The Antennae Galaxies are said to be interacting, a polite way of saying that they are colliding with each other. As they do so, gravitational forces trigger massive amounts of star formation revealed in long-exposure photographs as numerous clusters filled with hot, bright stars. Trailing behind the galaxies are two immensely long tails of dust and gas, and it is from these that the Antennae Galaxies get their name.

The Antennae Galaxies were believed to be 65 million light years away until recently, when astronomers revised their estimates, placing the galaxies a mere 45 million light years away. Because they are relatively close, these galaxies are of great importance to astronomers looking to understand how galaxies interact with one another, and what happens when they do. Galaxy collisions aren't common, but they do happen from time to time, and our own Milky Way Galaxy will probably collide with the Andromeda Galaxy in a few billion years.

Both galaxies are, as you'd expect, members of the same galaxy group, in this case the NGC 4038 Group. They are not the brightest members, though; that accolade goes to NGC 4027, a barred spiral galaxy also in Corvus.

NGC 4088

NGC 4088 is an intermediate spiral galaxy in the constellation of Ursa Major situated about 60 million light years away from Earth. It is reasonably bright but from our viewpoint at least has little by way of discernible features. Under good conditions, it appears to be little more than a dim, irregular smear of light.

Owners of wide-field Newtonian telescopes often place this galaxy in the same field of view as its neighbors NGC 4026 and NGC 4157. To do this with a go-to telescope you will need a field of view of around 2 degrees. With NGC 4088 in the center of the field, the other two galaxies should be towards the edge of the field of view, one on either side of NGC 4088.

NGC 4088 is a member of the Ursa Major North Group of galaxies, also known as the M109 Group.

NGC 4111

This small, fairly bright galaxy in Canes Venatici is a lenticular galaxy with a bright core that appears almost star-like at the eyepiece. It is about 60 million light years away, and is a member of the Ursa Major North Group of galaxies, sometimes called the M109 Group.

NGC 4157

NGC 4157 is a barred spiral galaxy in Ursa Major that we view edge-on, so appears to be little more than a thin sliver of light with a brighter core. Small telescopes may only show the core, in which case this object may look almost star-like, but telescopes with apertures upwards of 200 mm should allow you to see the entire object.

Wide-field instruments can place NGC 4157 in the same field of view as two other galaxies, NGC 4088 and NGC 4026. You will need a field of view of 2 degrees though. If you have your go-to telescope place NGC 4088 in the center of the field of view, the other two galaxies should be towards the edge of the field of view, one on either side of NGC 4088.

NGC 4088 is a member of the Ursa Major North Group of galaxies, also known as the M109 Group.

NGC 4192 (M98)

This relatively dim Messier object in Coma Berenices is a spiral galaxy but viewed edge-on, so it appears as little more than a thin wispy glow with a bright, almost star-like, core. Under dark skies large telescopes will reveal more detail, including a mottled surface and prominent dust lanes, but these won't be apparent to the average observer under suburban skies.

M98 is about 60 light years away and a member of the Virgo Cluster of galaxies. It is unusual among Virgo Cluster galaxies in moving towards the Milky Way Galaxy rather than away from it. Indeed, it is zooming towards us at more than 200 km per second! Still, given the vast distances involved, there's no risk of a collision any time soon. . .

NGC 4214

NGC 4214 is a reasonably bright irregular barred galaxy in the constellation of Canes Venatici. It is a dwarf galaxy somewhat like the Magellanic Clouds in terms of size and shape. Depending on your observing conditions, its core should be apparent as a somewhat elliptical patch of light. Dark skies and large apertures don't reveal much more detail, but the wispy halo around the core should be more obvious.

By galaxy standards NGC 4214 is quite nearby, a mere 13 million light years away. It is notable for containing numerous areas of intense star formation visible in long-exposure photographs as bright star clusters. Possibly this was triggered by a recent interaction with another galaxy. Clusters of older red giant stars reveal that there have been periods of intense star formation in the past too, and there have likely been bouts of star formation in NGC 4214 across the last couple of billion years.

NGC 4216

This barred spiral galaxy in Virgo is viewed almost edge-on, and appears distinctly spindle-like in shape through most telescopes. It has a bright core and only some-what less luminous arms.

NGC 4216 is about 65 million light years away and is a member of the Virgo Cluster. Unlike most other members of that cluster of galaxies, it is moving towards us rather than away.

NGC 4244

Like NGC 4216, this spiral galaxy in Canes Venatici is viewed edge-on, so appears as little more than long thin patch of light. It is fairly bright and easy to see, even under moderately dark suburban skies, and well worth tracking down. It is very long and thin, strikingly so, though its bright core should be apparent. Long-exposure photographs reveal numerous details including the fact stars appear to be clumped around the disc, presumably because from our viewpoint that's how we see the spiral arms.

NGC 4244 is the largest of the galaxies that make up the Canes I Group, measuring some 65,000 light years in diameter. Estimates on the distance of NGC 4244 from Earth vary from 4 to 14 million light years, but either way, it's one of the relatively nearby galaxies.

NGC 4254 (M99, Coma Pinwheel Galaxy)

Although a remarkable object in photographs, this galaxy is apt to disappoint at the eyepiece. The bright core should be visible readily enough, assuming fairly dark skies, but its most dramatic feature, its asymmetrical spiral arms, aren't usually apparent. What long-exposure photographs reveal is that one of the spiral arms is thicker and looser than the other, almost as if stretching out towards something. But what?

One hypothesis is that a so-called dark galaxy VIRGOHI21 is moving past M99 sufficiently closely that its gravity is pulling that spiral arm towards it. VIRGOHI21 certainly contains hydrogen, but astronomers believe it might also contain dark matter, and it's this stuff that makes VIRGOHI21 sufficiently massive that it can distort M99. As VIRGOHI21 and M99 drift past each other, the gravitational interaction will fade away, and M99's arm will return to its normal shape.

M99 is about 60 million light years away and a member of the Virgo Cluster, as is, presumably, VIRGOHI21.

NGC 4274

This barred spiral galaxy in Coma Berenices is faint and difficult to see under suburban conditions. It is viewed obliquely, and usually appears as a faint, elliptical smear of light. Long-exposure photographs shows a bright core with tightly coiled spiral arms, but such details are difficult to see at the eyepiece.

NGC 4274 is about 40 million light years away and the brightest member of a group of seventeen galaxies known as the NGC 4274 Group.

NGC 4303 (M61)

M61 is a spiral galaxy in the constellation of Virgo. It is viewed face-on, and in long-exposure photographs its spiral arms and bright central core are easily seen, as well some prominent dust lanes. At the eyepiece though these features are not easily seen, and small to medium sized telescopes will probably only allow the observer to see the core of the galaxy.

This galaxy is one of the larger members of the Virgo Cluster, of comparable size to our own Milky Way Galaxy. In 2008, a sixth supernova had been observed in this galaxy, more than for any other Messier catalog galaxy save M83, for which six had also been recorded. The absolute record though is held by NGC 6946, within which no fewer than nine had been observed by 2009.

NGC 4321 (M100)

This fairly bright spiral galaxy in Coma Berenices is yet another member of the Virgo Cluster. It is similar in size to the Milky Way Galaxy, containing over 100 billion stars, but like many other galaxies oriented towards us face-onwards, it isn't particularly easy to see. Realistically, you'll need at least a 100 mm aperture to see this galaxy, and likely 150 mm upwards under mediocre suburban skies. Even then, you'll probably only see the bright central core.

Long-exposure photographs show this galaxy to be exceptionally beautiful, and a classic example of a Grand Design spiral galaxy. It has sharply defined spiral arms crisscrossed with dust lanes and numerous clusters of young, hot, blue stars.

Astronomers have also spotted those variable stars known as Cepheid Variables in M100. These are useful for judging distances because the rate at which they vary is related to their absolute magnitude (i.e., how bright they actually are, as opposed to their apparent magnitude, which is how bright they seem from our distant vantage point). By comparing their absolute magnitude to their apparent magnitude, you can estimate how far away they are, because the more distant a star of given absolute magnitude, the fainter its apparent magnitude will be. In 1996 astronomers deduced what they believe is the first accurate measurement of the distance between us and a galaxy outside the Local Group, in this case establishing that M100 is about 56 million light years away, with a margin of error of about 6 million light years either way.

NGC 4374 (M84)

M84 is a lenticular galaxy in the constellation of Virgo, and one of the Virgo Cluster of galaxies. It's fairly bright and easy to spot, but apart from its overall shape, there's little to be seen. Long-exposure photographs show very prominent dust lanes, but these won't be visible at the eyepiece. At least two supernovae have been observed in this galaxy.

NGC 4382 (M85)

This lenticular galaxy is a remarkably unexciting object at the eyepiece. There's little to be seen, It's relatively bright though, and should be visible as an elliptical blur through apertures even as small as 90 mm under reasonably good suburban skies.

Indeed, under dark skies M85 can even be glimpsed with binoculars. A companion galaxy, NGC 4394 is very close by and through telescopes with apertures 150 mm upwards should be visible as a faint, hazy, but otherwise star-like object.

M85 is a member of the Virgo Cluster and lies about 60 million light years away.

NGC 4388

This spiral galaxy in Virgo is not bright and very easy to miss under suburban conditions. There's not much to see even if you can spot it, but astronomers have found this galaxy to be rather puzzling. Vast clouds of gas more than 100,000 light years in length are streaming out of the core, but precisely why, nobody knows. One theory is that NGC 4388 collided with another galaxy at some point, and the gas is all that remains of that other galaxy.

NGC 4388 is a member of the Virgo Cluster and lies about 60 million light years away from Earth.

NGC 4406 (M86)

Like many other lenticular galaxies, this galaxy in the constellation of Virgo is relatively easy to see but not particularly interesting to look at. There isn't much detail apparent, even with large-aperture instruments.

M86 is yet another member of the Virgo Cluster and lies about 52 million light years away. M86 is the Messier catalog object moving most quickly towards us, at about 244 km/s (over half a million miles per hour!).

NGC 4414

NGC 4414 is a spiral galaxy in Coma Berenices that falls into a particular class of object known as flocculent spiral galaxies. As this name suggests, instead of smooth spiral arms typical of Grand Design spiral galaxies, long-exposure photographs reveal flocculent spiral galaxies have a much more lumpy texture a bit like curdled milk. In effect, their arms are broken up into numerous small chunks.

The reasons why some spiral galaxies evolve into one form and others into the other remain unclear, but the mode of star formation is probably important. Grand Design spiral galaxies are dominated by star formation triggers by gravity waves, possibly initiated by interactions with neighboring galaxies. By contrast star formation in flocculent galaxies occurs on a more ad-hoc basis, the rotation of the galaxy sending clouds of dust and gas on the collision courses necessary to initiate star formation.

NGC 4414 is about 62 million light years away and a member of the Virgo Cluster.

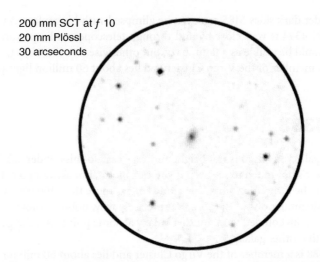

200 mm SCT at f 10
20 mm Plössl
30 arcseconds

Figure 3.14. NGC 4414 reveals an unusual lumpy texture when photographed, but through a 200-mm SCT it is nothing more than a faint elliptical blur. (Image produced using Starry Night Pro. AllSky data courtesy of Main-Sequence Software Inc.)

NGC 4438

NGC 4438 (a spiral galaxy) and NGC 4435 (a lenticular galaxy) form a pair of galaxies known as The Eyes. Under dark skies, they appear as two quite elliptical blurs of light that are quite close together. A 200-mm SCT equipped with a 20-mm Plössl should comfortably fit them into the same field of view, the two objects being separated by a mere 4 arcminutes. NGC 4438 is the brighter of the two objects.

Both galaxies are members o the Virgo Cluster and are located about 52 million light years away. The two galaxies are interacting with each other, and have likely also been disturbed by another galaxy as well, M86. Photographs have revealed huge tendrils of gas arcing the 400,000 light years distance between NGC 4438 and M86, and astronomers believe these imply the two galaxies collided with one another in the past.

NGC 4449 (Box Galaxy)

This irregular galaxy in Canes Venatici is apparently very similar to the Large Magellanic Cloud in size and composition. Through small telescopes it has a distinctly rectangular shape, and has consequently become known as the Box Galaxy. It is reasonably bright and easy to see under moderately dark suburban skies, and with an aperture of 200 mm the hazy area around its bright bar-shaped core should be apparent.

NGC 4449 is a member of the Canes Venatici I Group a mere 13 million light years away.

NGC 4501 (M88)

M88 is a spiral galaxy in Coma Berenices and a member of the Virgo Cluster. It is about 47 million light years away. It isn't particularly bright or easy to see, though it should be visible under dark skies through telescopes even as small as 90 mm. More realistically, under suburban skies you'll want an aperture of at least 150 mm, and larger telescopes should reveal a hazy region around the bright core; these are all you can see of its spiral arms.

NGC 4517

NGC 4517 is a faint spiral galaxy in Virgo that is viewed edge-on. It isn't easy to see and may be completely invisible through small aperture instruments or under light-polluted skies. If you can't see it, you aren't missing much: it's essentially a thin, elongated blur with no obvious brightening of the core. NGC 4517 is about 55 million light years away and a member of the Virgo Cluster.

NGC 4526

NGC 4526 is a barred lenticular galaxy in Virgo that appears as strongly elliptical blur with a noticeably bright core. It is a small and dim object, and not at all easy to see. Under suburban conditions, at least 150 mm, and likely 200 mm, will be required for a decent view. NGC 4526 is sometimes called the Lost Galaxy because it is so hazy and difficult to see. This galaxy is close to the core of the Virgo Cluster. It is about 55 million light years from Earth.

NGC 4535

NGC 4535 is a dim spiral galaxy in Virgo and another member of the Virgo Cluster. This galaxy is not far from NGC 4526, and large aperture, wide-field telescopes may show them in the same field of view under good, dark conditions. The two galaxies are about 35 arcminutes apart, placing them well within the 50 arcminute field of view delivered by a 32-mm Plössl when used with a standard 200-mm SCT.

NGC 4548 (M91)

This is another moderately bright barred spiral galaxy in Coma Berenices, but while its core is easily seen even with fairly small telescopes, only a large telescope will show its spiral arms. It is estimated to be about 63 million light years away from Earth, and is a member of the Virgo Cluster of galaxies. There is some debate as to whether this actually is the galaxy Messier recorded as object number 91 on his famous list, but it's the one normally referred to as M91 by amateur astronomers.

NGC 4552 (M89)

M89 and M90 (NGC 4569) are a pair of fairly bright galaxies in Virgo only a mere 40 arcminutes apart, and can consequently may be fitted into the field of view if a low-power (or wide-field) eyepiece is used. In the case of a standard f/10 200 mm-SCT, a 32-mm Plössl will do the trick nicely.

M89 is technically an elliptical galaxy, but it appears to be almost perfectly spherical. This might be simply because of the way the galaxy is oriented towards, in the same way that an egg looks circular if viewed end-on. But it might also be a genuinely spherical galaxy. M89 has an unusually large entourage of globular clusters, estimated to number around 2,000, something like ten times as many as those that orbit the Milky Way Galaxy.

NGC 4559

NGC 4559 is a barred spiral galaxy in Coma Berenices that is fairly bright and easy to see. It's a little smaller than the Milky Way Galaxy, at about 900,000 light years in diameter. NGC 4559 is a member of the Virgo Cluster and lies about 26 million light years away.

NGC 4567 and NGC 4568 (Siamese Twin or Butterfly Galaxies)

These two galaxies are members of the Virgo Cluster and lie about 50 million light years away. They are both spiral galaxies, though little of their structure can be seen at the eyepiece. At best, you'll see a faint, Y-shaped blur. What makes them interesting is that while they appear to be on top of each other, and are probably interacting, neither of them display any of the features typical of interacting galaxies, such as tidal tails or twisted spiral arms.

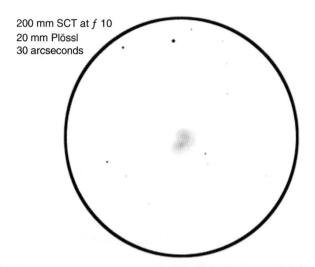

200 mm SCT at ƒ 10
20 mm Plössl
30 arcseconds

Figure 3.15. The Butterfly Galaxies are a pair of interacting galaxies that don't exhibit the distortions typical of interacting galaxies. (Image produced using Starry Night Pro. AllSky data courtesy of Main-Sequence Software Inc.)

NGC 4569 (M90)

M90 forms a neat pair of fairly bright galaxies with M89, the two being located in Virgo and easily seen in a low-power, wide-field eyepiece. A 32-mm Plössl used with a 200 mm-SCT should place the two galaxies in the same field of view, about 40 arcminutes apart. Both galaxies are bright enough to be seen under reasonably dark suburban skies.

M90 is about 58 million light years away, and unusual among Virgo Cluster galaxies in that it is moving towards Earth rather than away from it. Although a spiral galaxy, some astronomers believe that it is a very old galaxy gradually becoming more lenticular in shape. One peculiarity is its overall lack of areas with active star formation.

NGC 4594 (M104, Sombrero Galaxy)

This famous galaxy in Virgo is a spiral galaxy viewed almost, but not quite, edge-on. It has a prominent dust lane around its perimeter, but because of the angle at which we view this galaxy, it seems to be divided into two unequal halves. The result is, in long-exposure photographs at least, what appears to be something like a sombrero hat sitting on a round head.

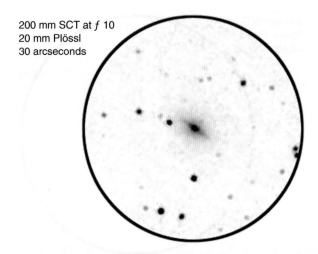

200 mm SCT at f 10
20 mm Plössl
30 arcseconds

Figure 3.16. M104 is a favorite target for astronomers under dark skies, but while its core is easily seen under suburban skies, its prominent dust lane will be much more difficult to see. (Image produced using Starry Night Pro. AllSky data courtesy of Main-Sequence Software Inc.)

The Sombrero Galaxy is fairly bright and easily visible under dark skies through small telescopes. Large telescopes, from 200 mm upwards, should reveal the dust lane. The Sombrero Galaxy is a member of the Virgo Cluster and about 50 million light years away.

NGC 4605

NGC 4605 is a dwarf spiral galaxy in Ursa Major about the same size as the Large Magellanic Cloud. It has a bright core but little else is visible through telescopes even as large as 200 mm in aperture. It is about 15 million light year away.

NGC 4621 (M59)

This elliptical galaxy in Virgo is only 24 arcminutes away from M60 (NGC 4649) and consequently the two galaxies are often viewed together. A low-power eyepiece works best, a 32-mm Plössl for example easily placing them in the same field of view when used with a 200-mm SCT. While you can't see much detail, these two galaxies are at least reasonably bright and should be plainly visible under fairly dark skies.

M59 is about 60 million light years away and a member of the Virgo Cluster.

NGC 4631 (Whale Galaxy)

This edge-on spiral galaxy in Canes Venatici is quite bright and should be visible under reasonably dark skies as a long, thin streak. It is interacting with a much smaller companion galaxy, NGC 4627, resulting in a distorted shape likened to that of whale. Another large galaxy, NGC 4656, is about 32 arcminutes away, and if you use a low power eyepiece, you should be able to get them both into the same field of view. On a 200-mm SCT, a 32-mm Plössl works well.

NGC 4631 is about 30 million light years away and is a member of the NGC 4631 Group of galaxies.

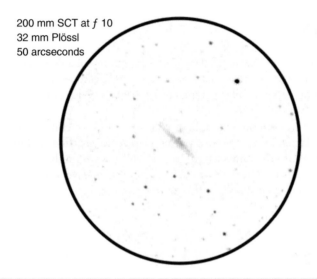

200 mm SCT at ƒ 10
32 mm Plössl
50 arcseconds

Figure 3.17. The Whale Galaxy is a spiral galaxy viewed edge-on. Although essentially spindle-shaped, one end is twisted upwards by the gravitational pull of a neighboring galaxy. The result is something that looks like a whale with its flukes lifted upwards, as if preparing to dive. (Image produced using Starry Night Pro. AllSky data courtesy of Main-Sequence Software Inc.)

NGC 4649 (M60)

M60 is a fairly bright elliptical galaxy in the constellation of Virgo. As mentioned above, it is quite nearby M59 (NGC 4621) and the two objects can be easily seen in the same field of view if a low-power eyepiece is used. M60 is about 55 million light years away and a member of the Virgo Cluster.

A small, faint galaxy, NGC 4647, may be visible close to M60 under dark skies. It is apparently interacting with M60

NGC 4656 and NGC 4657 (Hockey Stick Galaxies)

This pair of interacting galaxies are members of the NGC 4631 Group of galaxies and located about 25 million light years away. Their common name comes from how they appear at the eyepiece when viewed through a reasonably large telescope under dark skies. The bigger and brighter of the two galaxies, NGC 4656 forms the handle, and the smaller and dimmer galaxy NGC 4657 forms the curved head.

NGC 4699

This spiral galaxy in the constellation of Virgo is reasonably bright and the core should be easily seen under dark skies through telescopes of aperture 150 mm upwards. Long-exposure photographs reveal delicate spiral arms, but these usually aren't visible at the eyepiece.

NGC 4699 is about 60 million light years away and a member of the huge Virgo Cluster of galaxies. In fact the galaxy gives its name to a particular clump of galaxies, the NGC 4699 Group, numbering at least 12 galaxies within the southern extension of the Virgo Cluster. The NGC 4699 Group is one of several such groupings that astronomers classify as the Virgo II Groups.

NGC 4725

NGC 4725 is a barred spiral galaxy in Coma Berenices. One of the peculiar things about this galaxy is that it only has one arm; most spiral galaxies have at least two. This feature isn't visible through the eyepiece though, where all that is visible is the bright core surrounded by a certain amount of haziness. This galaxy is about 40 million light years away.

NGC 4762

This lenticular galaxy in Virgo lies between three foreground stars, two on one side, and one on the other. It is a fairly bright galaxy and should be visible under good conditions in telescopes 150 mm in aperture or even less. There isn't much to see, but its streak-like shape should be obvious. The central bulge of the core is obvious in photographs, but much less so visually. NGC 4762 is about 67 million light years away and a member of the Virgo Cluster.

NGC 4889

This elliptical galaxy in Coma Berenices is one a vast number of galaxies swarming around this particular part of the sky not far from the North Galactic Pole. Most of these galaxies, including NGC 4889, belong to the Coma Cluster, a grouping of more than a thousand galaxies that, together with the Leo Cluster, make up the huge Coma Supercluster.

Amateurs with large aperture instruments and pitch black skies can spend a lot of time surfing from one galaxy to the next, but those under suburban skies will find things a lot more difficult. Although NGC 4889 is the brightest of the Coma Cluster galaxies, it's still a pretty dim object. It is about 250 million light years away, making it a far more distant object than many of the other galaxies viewed by amateur astronomers.

NGC 5005 and NGC 5033

These two galaxies form a pair that are sufficiently close together that their gravitational fields affect one another, but are not so close together that their shapes have been distorted. Both are spiral galaxies, NGC 5005 being a little brighter and easier to spot, but neither showing much at the eyepiece beyond a faint elliptical blur. NGC 5033 is a Seyfert Galaxy and believed to contain a supermassive black hole at its center.

A low-power or wide-field eyepiece should be able to fit both galaxies in the same field of view since the separation between them is only about 42 arcminutes. On a 200-mm SCT, a 32-mm Plössl should work fine.

NGC 5746

NGC 5746 is a reasonably bright spiral galaxy in Virgo that is viewed edge-on. Under dark skies it is easily visible as a long, thin blur of light. Long-exposure photographs reveal a slight central bulge and a prominent dust lane around the perimeter of the galaxy. NGC 5746 is 90 million light years away and a member of the Virgo-Libra Cloud of galaxies.

NGC 5866 (M102, Spindle Galaxy)

This fairly bright lenticular galaxy in Draco is sometimes known as the Spindle Galaxy, though note that NGC 3115 in Sextans is also known by this name as well. When it comes to its inclusion in the Messier catalog, while NGC 5866 is often assumed to be M102, there is a bit of debate over whether that is actually the case. A number of other galaxies have been proposed, and some astronomers even maintain that M101 and M102 are the same object that happened to be listed twice, presumably by accident.

In any case, NGC 5866 is unusual among lenticular galaxies in having what appears to be a ring of dust around its perimeter. Such features are more commonly seen on spiral galaxies, and one explanation for this is that NGC 5866 is simply a spiral galaxy that happens to look like a lenticular galaxy from our particular perspective.

NGC 5866 is a member of its own small group of galaxies, the NGC 5866 Group, and is located about 50 million light years away. Some astronomers believe that the NGC 5866 Group belongs to a larger grouping that also includes the M51 Group and the M101 Group.

NGC 5897

This fairly bright globular cluster in Libra is easy to see with telescopes as small as 100 mm in aperture, but an aperture of at least 150 mm will be required to begin resolving it. A larger aperture telescope, coupled with a reasonably high magnification, should help resolve the outer layers of stars. It's a very loose globular cluster, rated as a Class XI globular cluster on the Shapley–Sawyer scale that runs from I to XII. Like many objects in Libra, the quality of the views northern observers will get of this object will depend on how clear (and dark) their southern horizon happens to be.

NGC 5897 is about 40,000 light years away.

NGC 5907

NGC 5907 is a remarkable spiral galaxy in Draco that appears to have swallowed up a dwarf galaxy at some point, leaving colossal whorls of dust and gas around itself. These trails wind outwards some 150,000 light years from the core of NGC 5907. These amazing features are only visible on long-exposure photographs though, and amateur astronomers will only see this edge-on oriented galaxy as a faint and thin strip of light.

NGC 5907 is about 39 million light years away and a member of the NGC 5866 Group of galaxies.

NGC 6503 (Lost-in-Space Galaxy)

NGC 6503 is a small spiral galaxy in Draco lying about 17 million light years away. It lies in a bit of the sky known as the Local Void, an area between a number of major galaxy clusters but itself largely devoid of galaxies. NGC 6503 is an exception, and noted observer Stephen James O'Meara has referred to this galaxy as the Lost-in-Space Galaxy because of this.

NGC 6503 is reasonably bright and easy to see, though you will need a telescope of moderate to large aperture to do so. It is a member of the Pegasus Spur group of galaxies.

Colorful and Curious Stars

SAO 308 (Alpha Ursae Minoris)

Polaris is well known as the Pole Star, but it is also a challenging double star for small aperture telescopes. The primary is a yellow, slightly variable supergiant star with a magnitude 9.0 companion. The separation between the two stars is just over 18 arcseconds.

SAO 8024 (5 Ursae Minoris)

There are three components to this star. Splitting magnitude 9.8 component C from the magnitude 4.3 primary A is easy, since there's almost an arcminute of space between them. Component B is much more difficult. It is a magnitude 13.3 star just over 13 arcseconds away from the primary. While that distance sounds quite generous, the huge difference in brightness between the two stars makes C very difficult to detect. A challenge for large aperture telescopes only.

SAO 15274 (VY Ursae Majoris)

This irregularly variable star is one of the reddest in the sky. It is a carbon star, like many other unusually red stars, and its brightness drops as clouds of carbon soot appear on its surface, only to brighten up once more as the soot dissipates. At its brightest this is a magnitude 5.9 star, dropping to magnitude 7.0 at its dimmest.

SAO 16273 (Alpha Draconis, Thuban)

At magnitude 3.7, Alpha Draconis is easily overlooked, but it is of considerable importance in archaeoastronomy. Because Earth 'wobbles' slightly as it rotates, the north–south axis doesn't always point in the same direction. At the moment it points towards Polaris, what we often call the Pole Star. But if you were to travel back to around 2800 BC, the Pole Star wouldn't be Polaris but Alpha Draconis instead. In fact it Alpha Draconis would be a mere 5 arcminutes away from the celestial pole; today, Polaris is over 40 arcminutes away from the celestial pole.

Because it was the Pole Star at the time, Alpha Draconis might have been significant to ancient civilizations. Inevitably, perhaps, ideas have sprung up connecting the star to the pyramids of Egypt. In particular, a small vent-like shaft running from the King's Chamber in the Great Pyramid at Giza seems to have been aligned in such a way that it points towards Alpha Draconis. Precisely what function it might have had, if any, remains controversial.

SAO 27876 (Beta Ursae Majoris, Merak)

Beta Ursae Majoris is the brightest star in the Ursa Major Moving Group, a collection of stars apparently heading away from the same point in space. Since all the stars in this collection have similar ages and composition, the assumption is that they were once members of a star cluster that was formed from a single nebula some 500 million years ago. Over time the stars drifted apart, and the star cluster they once formed is now undetectable.

Quite a few of the brighter stars in Ursa Major are members of this group as well, including Beta Ursae Majoris (Merak), Gamma Ursae Majoris (Phecda), Delta Ursae Majoris (Megrez), Epsilon Ursae Majoris (Alioth), and both members of Zeta Ursae Majoris (Mizar and Alcor).

Perhaps more surprisingly, there are also numerous members of the Ursa Major Moving Group that are members of constellations other than Ursa Major, including Delta Aquarii, Beta Aurigae, Zeta Boötis, Chi Ceti, Zeta Crateris, Alpha Coronae Borealis, Gamma Leporis, Gamma Microscopii, Chi-1 Orionis, Beta and Omega Serpentis, and Zeta Trianguli Australis.

All told, there are at least 60 stars in the group, and possibly as many as 220. The cluster covers a huge amount of space, but the core is only about 80 million light years from Earth. However, at least one traditionally included member, Sirius, is now known not to be a member, and is probably a member of an entirely different group, referred to as the Sirius Supercluster.

By itself Merak might not be a very interesting star, but the Ursa Major Moving Group does at least provide a fascinating glimpse at the ultimate fate of open star clusters like M35, the Pleiades and the Hyades.

SAO 28737 (Zeta Ursae Majoris, Mizar and Alcor)

Zeta Ursae Majoris is a double star that can be split with the naked eye. Mizar is a magnitude 4.0 white star, and 12 arcseconds away is Alcor, a magnitude 2.2 star of similar color. Although they are physically quite close to each other, Mizar 81 light years away from us and Alcor 78 light years away, they aren't orbiting one another and don't form a binary system. That said, anyone living on a planet orbiting Alcor would have an incredible view of Mizar shining with Venus-like brilliance in the night sky.

A thousand years ago these two stars were together known as the Puzzle, *al Sadak*, to Arabs and Persian astronomers. They insisted that this double star was a good test of visual acuity. Islamic scholarship in the fields of both astronomy and ophthalmology were second to none at the time, but the idea that Mizar and Alcor would make a useful test of a person's eyesight is highly questionable. Modern day astronomers such as Guy Consolmagno and Sir Patrick Moore have pointed out that the test is a very easy one: unless you have particularly poor eyesight, splitting them is not at all difficult.

Moore has suggested that perhaps the test involved the third star in the system, a blue–white magnitude 8.8 star known as Sidus Ludoviciana roughly halfway between Mizar and Alcor. Its name means Ludwig's Star and honors Ludwig V, Landgrave of Hesse (though somehow the name of the politically astute German astronomer who named the star after Ludwig has been lost). The problem is that as a magnitude 8.8 star, Ludwig's Star is too dim to be seen with the naked eye. Was Ludwig's Star brighter in the past? Possibly, but Mizar has been systematically observed for hundreds of years and photographed thousands of times, and not a trace of variability or flare star behavior has ever been detected.

Mizar is itself a double and historically notable as the first double star detected using a telescope. The magnitude 2.4 and 4.0 components are 14 arcseconds apart and easily split even with a small aperture telescope.

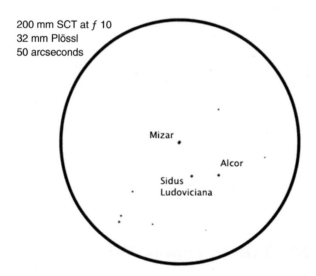

Figure 3.18. Mizar forms one corner of a triangular arrangement of stars. Alcor is the star at the corner furthest away, and Sidus Ludoviciana is the star about halfway between the two of them and off to one side. (Image produced using Starry Night Pro. AllSky data courtesy of Main-Sequence Software Inc.)

SAO 42876 (41 Lyncis)

Despite its name, 41 Lyncis is actually in the constellation of Ursa Major. A magnitude 5.4 orangey-yellow giant with a magnitude 8.0 companion about 1.4 arcminutes away. The primary star is of some scientific importance, being the star around which a planet 2.7 times larger than Jupiter was discovered in 2008. This planet is now known as 41 Lyncis b.

SAO 44317 (Y Canum Venaticorum, La Superba)

This beautiful, deep red star was given the name La Superba by the astronomer and Jesuit priest Angelo Secchi of the Pontifical Gregorian University. Secchi was one of the pioneers of spectroscopy, a technique that allows the spectrum of light emitted by stars to be measured and recorded. Once it was established that different types of stars emitted different spectra, and that this was related to their temperature, spectroscopy became the key tool for classifying stars. Using spectroscopy, Secchi was able to demonstrate that the Sun produced a similar spectrum of light to that of stars, proving for the first time what many scientists had hypothesized: that the Sun was a star.

The best way to appreciate its color is through binoculars or small aperture telescope; on large aperture telescopes, the view through the finderscope may well be more satisfying. Like many other deep red stars, La Superba is a carbon star, a giant star fusing helium into carbon, and consequently with a lot of soot in its atmosphere. It is the soot that gives it its peculiar tint. But La Superba is also a member of an unusual subtype known as J-type carbon stars, stars with unusually high quantities of the rare carbon isotope carbon-13. La Superba is a very large star, with a radius of more than 200 times of the Sun. If it was where the Sun is, its outer edge would be well beyond the orbit of Mars.

La Superba is a variable star, varying between magnitudes 4.8- and 6.3- on a 160-day cycle.

SAO 62053 (Beta Leonis Minoris)

Beta Leo Minoris is a notoriously difficult double star to split. It consists of a magnitude 4.4 primary and magnitude 6.1 secondary. The stars orbit each other in 38.6 years, and at their closest they are essentially inseparable. At the time of writing this they are about 0.5 arcseconds apart and will be getting further apart (and therefore easier to split) over the next few years.

SAO 63257 (Alpha Canum Venaticorum, Cor Caroli)

This double star is an easy split for telescopes of all sizes. Both components are blue–white, one at magnitude 2.9, the other at magnitude 5.6. The brightest of the two is interesting from a scientific point of view because it is a good example of what are known as magnetic variable stars. Its variation isn't enough to be seen visually, but astronomers have established that it has an unusually strong magnetic field as well as peculiarly high quantities of elements like silicon and europium compared with most other stars.

The name Cor Caroli is Latin for 'Charles' heart.' The court physician Sir Charles Scarborough said that he had observed that this star shone particularly brightly on May 29, 1660, when Charles II returned to London after his long exile abroad following the English Civil Wars. Initially the star was referred to as the heart of Charles the Martyr, meaning that it was actually the recently executed Charles I who was the one being honored. But because of the connection with the Restoration and Charles II, it has sometimes been assumed that the star's name honors the second King Charles.

SAO 64686 (Mu Boötis, Alkalurops)

Alkalurops is good triple-star system for small telescopes. It consists of a yellow F-type star at magnitude 4.3 (SAO 64686) and two orangey G-type stars at magnitudes 7.2 and 7.8 (together counted as SAO 64687). The latter two stars, the B and C components, are separated from each other by only 2.2 arcseconds and can be split with telescopes as small as 75 mm in aperture, but this does require high magnification and steady skies.

SAO 82279 (12 Comae Berenices)

12 Comae Berenices is one of the brighter stars in the cluster known as Melotte 111 or the Coma Star Cluster. Spanning more than 5°, this is object is a classic target for binoculars and rich field telescopes.

SAO 81298 (Gamma Leonis, Algieba)

This is a very highly regarded double star consisting of an reddish-orange magnitude 2.2 primary and a yellow magnitude 3.5 secondary. The two stars are separated by 4.4 arcseconds. The secondary star often appears green because of the contrast between its color and that of the primary. There are no genuinely green stars, but this is one of the very few that appears to be green, at least to some observers.

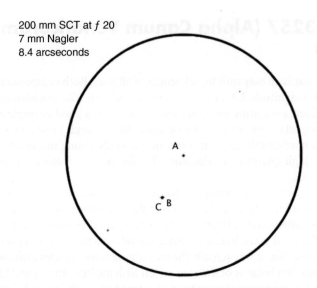

200 mm SCT at f 20
7 mm Nagler
8.4 arcseconds

Figure 3.19. Alkalurops consists of a fourth-magnitude component and two very seventh-magnitude components separated from each other by 2.2 arcseconds. While the system can be resolved with a small telescope, a steady sky and high magnification will be required. (Image produced using Starry Night Pro. AllSky data courtesy of Main-Sequence Software Inc.)

SAO 81583 (54 Leonis)

This is an attractive double star consisting two blue–white of magnitudes 4.5 and 6.3, separated by 6.6 arcseconds. An easy split, even for small telescopes.

SAO 83500 (Epsilon Boötis, Izar)

Epsilon Boötis is an attractive double star that is notoriously difficult to split with small telescopes. It's a good test of the quality of your telescope and eyepieces. Good quality refractors can split this double with apertures upwards of 76 mm, but telescopes with central obstructions, such as reflectors, may need somewhat more aperture for a clean, obvious split. The primary is a magnitude 2.5 orange-colored K-type giant star, while the secondary is a blue, magnitude 4.6 A-type main sequence star. Some observers find that the dimmer of the two stars appears green rather than blue.

SAO 84015 (R Coronae Borealis)

R Coronae Borealis is the prototype of a class of variable star known as R Coronae Borealis type stars (often abbreviated to R CrB type stars). Mostly this star shines at magnitude 5.8, but periodically, and unpredictably, its brightness plummets down to 14.8 across just a few weeks. It them takes months for it to regain its original brightness. The mechanisms behind this variability are obscure, but are believed to be related to the production of clouds of carbon soot on the surface of the star and then their subsequent dispersal. As the soot comes and goes, the star's brightness falls and rises.

SAO 100160 (24 Comae Berenices)

This double star is merely a chance alignment of two stars, the brighter (magnitude 5.1) orange star about 615 light years away, and the dimmer (magnitude 6.3) white star over 2,700 light years away. The two stars are separated by just over 20 arcseconds, so they're an easy split, even with small aperture telescopes.

What does make this pair memorable is the color contrast between the two. As is occasionally the case where two differently colored stars are viewed close together, the brain sometimes fails to detect their correct colors. So rather than the dimmer star being perceived as a white star, many observers find that it seems to be green.

SAO 101624 (Delta Serpentis)

Delta Serpentis consists of two white stars, one at magnitude 4.2 and the other at magnitude 5.2, separated by 4.4 arcseconds. Because of the considerable difference in brightness this pair is moderately difficult to split in small aperture telescopes, but certainly doable given steady skies and high magnification.

SAO 117717 (Omega Leonis)

Omega Leonis is a tough double star for large aperture telescopes. The two components are separated by 0.6 arcseconds, so while similar in brightness (magnitudes 5.9 and 6.5) you will need good conditions and high magnifications to tease them apart.

SAO 124070 (Theta Serpentis)

Theta Serpentis consists of a pair of very similar A-type main sequence stars, both white in color. Some older books describe the stars as both having magnitudes around 4.5, but more recent estimates place their brightness at magnitudes 4.6 and 5.0. Possibly one of the stars is slightly variable. They are quite far apart, 22 arcseconds, and easily split even with a small telescope.

SAO 123778 (and IC 4756)

SAO 123778 is an outlying member of the open star cluster IC 4756. This cluster is almost a degree in diameter, and generally considered a suitable target for astronomers using binoculars or rich field telescopes. In fact it can be spotted with the naked eye under dark sky conditions. When using an SCT, be sure to use a reducer-corrector and the lowest-power, widest-field eyepiece you have. IC 4756 is about 1,300 light years away.

SAO 138917 (Gamma Virginis, Porrima)

Porrima is one of the brightest stars in Virgo, but notorious among double star observers for being difficult to split some years, and very easy on others. The explanation of course is that the secondary orbits the primary along an elliptical orbit, and this orbit is angled in such a way to us here on Earth that the secondary may be as far as 6 arcseconds from the primary, or as little as 0.3 arcseconds. The last time the two stars were at their farthest was in 1919, and were afterwards at their closest in 2005. From the early 1990s onwards this double star has been exceedingly difficult to split, and Porrima won't return to being an 'easy' double star until around 2020. The entire orbit takes about 170 years, though there's some debate over the precise value.

The two stars that make up this double shine at magnitude 3.5 and are white in color. You'll need at least 250 mm to split them, though smaller apertures may reveal an elongate rather than circular shape to the star, hinting at its binary nature if not clearly exposing it.

SAO 140030 (Beta Librae, Zubeneschamali)

There are no truly green stars. Even though stars emit green light, none emit enough green light to convince our eyes that the star is green in color. Whatever green light they're emitting is overwhelmed by the signals generated by our eyes by the red and blue wavelengths of light. By the time our brain has processed all the different signals together, we end up seeing white, blue, orange or red stars instead.

Very occasionally, double stars trick our eyes into seeing green. The contrast between the yellow and blue components is such the blue star can appear to be green. Epsilon Boötis (SAO 83500) and Gamma Leonis (SAO 81298) are the classic examples of this type of optical illusion in the springtime sky. The magnitude 2.6 blue dwarf star Beta Librae is even more exceptional: it is a solitary star that appears green, at least to some observers.

There's no good explanation for why Beta Librae appears to be green. Some people don't even see the greenness at all, and as ever in visual astronomy, the observer has to be careful not to see something he or she believes to be there. One idea is that this is in fact a double star system viewed edge on, with the light from a bluish star and the light from a more yellowy star somehow mixing in such a way that we see green.

SAO 156661 (Gamma Crateris)

The constellation of Crater doesn't contain much to amuse the owner of small to medium aperture telescopes, but the star Gamma Crateris is worth a look. It's a double star system with a magnitude 4.1 primary and magnitude 9.6 companion, the two stars a little over 5 arcseconds apart. Given good seeing conditions, they can be easily split with apertures upwards of 75 mm.

SAO 157323 (Delta Corvi)

When naming stars, Johann Bayer usually called the brightest star the alpha star, the next brightest beta, and so on. While there are a exceptions to this rule, no other constellation ignores the basic pattern as flagrantly as Corvus! Not only is Alpha Corvi dimmer than either Beta or Gamma Corvi, it is dimmer even than Delta Corvi!

In any event, Delta Corvi is a good double star for small aperture telescopes, its two components being separated by a respectable 24 arcseconds. The brighter of the two stars is a magnitude 2.9 blue–white star, while its companion is a magnitude 8.5 orange star. The color contrast between the two stars is notable, and the secondary star is often described as being purple or lilac rather than orange.

CHAPTER FOUR

Summer

The summer sky is dominated by the Milky Way, and for northern observers, this is the best time of the year to see that part of the Milky Way that passes through the southern constellations of Scutum, Sagittarius and Scorpius. What we're looking at is in fact the Sagittarius-Carina Arm of the Milky Way Galaxy. Moreover, the bit of this arm that we can see is closer to the galactic core than we are, and when we turn our telescopes towards Sagittarius, what we're actually doing is looking straight into the core of the Milky Way Galaxy.

Not only do we see lots of nearby nebulae and open star clusters when we study this part of the Milky Way, but we also see lots of globular clusters as well, since most of them orbit the galactic core. For northern sky observers summertime brings two of the best globular clusters into the night sky, M13 (NGC 6205) in Hercules and M22 (NGC 6656) in Sagittarius. While those two are the most celebrated, there are lots of others.

There are also lots of nebulae to be see as well. Among the showpiece summertime planetary nebulae are the Ring Nebula (NGC 6720) and the Dumbbell Nebula (NGC 6853). Suburban astronomers can expect to get good views of these objects, particularly if light pollution filters are used.

Good open star clusters abound, and being relatively unaffected by low levels of light pollution, they're great objects for the suburban astronomer. The best of them are in Scutum, Sagittarius and Scorpius, but since those constellations tend to hug the southern horizon, a good clear view of that horizon is critical.

It isn't all easy going at this time of year though! For northern hemisphere observers these will be the shortest nights of the year. That means it doesn't get truly dark until relatively late at night. Indeed, for observers as far north as England, it

N. Monks, *Go-To Telescopes Under Suburban Skies*, Patrick Moore's Practical
Astronomy Series, DOI 10.1007/978-1-4419-6851-7_4,
© Springer Science+Business Media, LLC 2010

may only be dark for an hour or two either side of midnight. Summer can also be rather 'buggy,' and unless you pack some mosquito repellent, you're likely find to find observing outdoors a bit of a trial once the bugs starting biting.

Showpiece Objects

NGC 6093 (M80)

This globular cluster in Scorpius is relatively small in angular size, a little over 7 arcminutes in diameter, but it is bright and easy to see, even with binoculars. It is only 3.5° away from another globular cluster, M4 (NGC 6121), and with binoculars it shouldn't be difficult to see the two objects in the same field of view, each one resembling a small but fuzzy patch of light. Of the two, M4 is the brighter and perhaps more interesting, but by any standards M80 is one of the better globular clusters and a good target for astronomers with small to medium aperture instruments.

Under good conditions the granular texture of M80 will be obvious even through a small aperture telescope, but this is a very dense globular cluster, and a large aperture telescope is required to resolve it down to the core. Unfortunately for observers at far northern latitudes, like many objects in Scorpius, getting to see M80 under good conditions will be difficult. Observers as far north as southern England will find that it rises little more than 15° above the horizon, so this is one object best viewed from a rural rather than built-up environment. Besides trees and buildings getting in the way, the dusty atmosphere above cities tends to make getting good sharp views of low-lying objects very difficult.

M80 is of middling distance from us at about 32,000 light years away and about 50 light years in diameter. It's rated as a Class II globular cluster on the Shapley–Sawyer scale. Among professional astronomers it is noted for being one of only two globular clusters within which a nova has been observed. It also has a relatively high proportion of stars known as blue stragglers, young stars that are hotter than the stars typically found in globular clusters. Precisely how blue stragglers evolve is a bit of a mystery, but it is thought that the unusually high density of stars in the core means that stars periodically collide with one another. This in turn creates the conditions needed for new stars to form.

NGC 6121 (M4, Cat's Eye Globular Cluster)

At a mere 7,200 light years away, the globular cluster M4 in Scorpius is one of the closest of its kind to Earth. As you'd expect, it is a bright and easily observed object when well placed in the sky, but far northern observers will be frustrated by its limited elevation above the horizon; at the latitude of southern England, it barely gets to 12° above the southern horizon. Observers farther south won't have too many problems though, and from southern Europe and most of the United States this is a superb object that hobbyists will want to look at again and again.

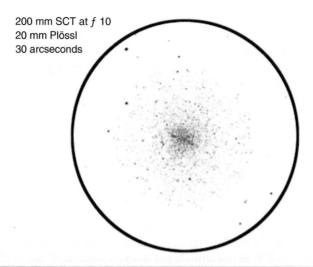

200 mm SCT at *f* 10
20 mm Plössl
30 arcseconds

Figure 4.1. M4 is a nice bright globular cluster with a distinctive band of bright stars running through the core. (Image produced using Starry Night Pro. AllSky data courtesy of Main-Sequence Software Inc.)

One quirky aspect to M4 is its proximity to the very bright red giant Antares, star being separated from the globular cluster by less than 1.5°. The two objects can be viewed together in binoculars with ease, but to the naked eye M4 is generally invisible, even though it should be bright enough to see.

Medium to large aperture telescopes (from about 150 mm upwards) should reveal another quirk. While most globular clusters contain stars that appear more or less even distributed, M4 has a very definite band of stars running along its center. As a result, some observers refer to this globular cluster as being the Cat's Eye Globular Cluster.

M4 is rated as a Class IX globular cluster on the Shapley–Sawyer scale, so it's relatively loose and looks very nice in medium to large aperture telescopes, the outer whorls of stars being comparatively easy to resolve. It is believed to be about 100 light years in diameter.

NGC 6205 (M13, Great Globular Cluster)

As the name Great Globular Cluster suggests, M13 is one of the finest globular clusters in the sky. This is because it is both intrinsically large (over 160 light years in diameter) and quite nearby (only 25,000 light years away). The net result is that it has a relatively large angular diameter of about 36 arcminutes as well as considerable brightness. At magnitude 7, it's just about visible to the naked eye under good, dark skies.

The shape and granular texture of M13 will be visible even through a fairly small telescope, but apertures upwards of 150 mm are needed if you want to resolve more stars that just the outermost whorls. Large apertures, from 200 mm upwards, pay dividends here, since middling to high magnifications are helpful. Since raising the magnification dims the image, the more aperture you have, the less detrimental this overall dimming of the image will be. Still, even with a 200-mm SCT, you'll find averted vision really helps.

Whether M13 is the fourth-best or third-best globular cluster depends on where you are. Observers and mid to far northern locations will find M13 rather better than M22 (NGC 6656) in Sagittarius, simply because M22 never rises very far above the horizon. By contrast M13 passes far overhead, so it's viewed through less of the dust and turbulence that affects objects viewed along the horizon. If these are the third and fourth best globular clusters, which are the best two? The best is probably Omega Centauri (NGC 5139) closely followed by 47 Tucanae (NGC 104). Neither is visible from mid to far northern latitudes, though Omega Centauri at least can be seen reasonably well from low latitude northern locations such as Florida.

M13 is about 12 billion years old and contains something like a half a million stars. At its core, individual stars are separated by less than ten astronomical units (where one astronomical unit is the distance between Earth and the Sun). In other words, the stars at the center of M13 are closer to each other than Saturn is to the Sun! It's rated at a middling Class V on the Shapley–Swayer, a nice balance between being compact enough to look good but loose enough to resolve easily. All in all, it's a deep sky object that most amateur astronomers enjoy observing whenever they get the chance.

In 1974, the famous Arecibo message was broadcast by the giant Arecibo radio telescope in the direction of M13. The message contained information, in binary code, that when read correctly would reveal some basic information about Earth and humanity. The point to choosing M13 was that it was conveniently placed in the sky at the time, and being a globular cluster, contained lots of stars in a very small chunk of space. While often described as an attempt to contact extraterrestrials, the Arecibo message was more of a technical exercise than anything else. Since M13 is orbiting the Milky Way core, it won't be in the same place after the 25,000 years it takes the Arecibo message radio signal to travel the 25,000 light year distance! So rather than a serious attempt at interstellar communication, the Arecibo message was more about showing how messages might be encoded, and what the required technology might be to broadcast them.

NGC 6218 (M12, Gumball Nebula)

The constellation of Ophiuchus contains several memorable globular clusters including M12. It is bright enough to be seen with telescopes of all sizes, and can be glimpsed with binoculars under sufficiently dark skies. Like most globular clusters it looks best through telescopes with at least 200 mm aperture, in part because

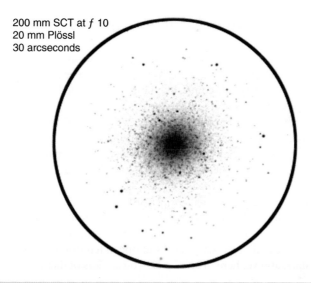

200 mm SCT at *f* 10
20 mm Plössl
30 arcseconds

Figure 4.2. The Great Globular Cluster in Hercules is the best globular cluster easily accessible to mid to far northern observers. Although there are globular clusters that are either as good (as in the case of M22) or better (Omega Centauri and 47 Tucanae), they're all southern sky objects. (Image produced using Starry Night Pro. AllSky data courtesy of Main-Sequence Software Inc.)

relatively high magnifications are needed to resolve anything more than the outer-most stars.

M12 is an interesting globular cluster to contrast with another one in Ophi-uchus, M10 (NGC 6254), a mere 3° away. Both are relatively nearby, M12 being about 16,000 light years distant and M10 a little over 14,000 light years distant, but the two clusters look very different. M10 is more compact and looks brighter. On the Shapley–Sawyer scale its rated at Class VII. By contrast, M12 is dimmer and seems more spread out, and on the Shapley–Sawyer scale its rated as a Class IX object.

M12 is about 75 light years in diameter and some 16,000 light years away. M12 is an odd cluster, possessing rather fewer low-mass stars than most other globu-lar clusters. The cluster contains about 200,000 stars, but astronomers believe that's about one-fifth the number it started out with, the other stars, mostly low-mass stars, having been stripped away, presumably by the gravitational pull of the Milky Way Galaxy.

Because the stars M12 was left with are the type that use up their energy reserves more quickly than low-mass stars, M12 won't last as long as most other globular clusters. So while most of the globular clusters orbiting the Milky Way will probably be around for another 20 billion years, M12 likely only has another 4.5 billion years of life left. While that's a small number compared to globular cluster averages, that's still a huge length of time, roughly equal to the age of Earth!

NGC 6231 (False Comet Cluster)

NGC 6231 is a nice but often overlooked open cluster in Scorpius. It isn't a very distant object, only about 5,900 light years away, and even a small telescope will reveal this cluster as a small, comet-like blur of light. Medium to large apertures easily resolve this cluster into its brightest members.

Unfortunately for northern sky observers, NGC 6231 is well below the celestial equator and only visible from low northern latitudes.

NGC 6231 is a very young object, some 3.2 million years, making it much younger than the Pleiades, which are about 100 million years old. It contains about 93 stars including numerous very large and very hot Class O stars. These are very obvious stars in this cluster, being among its brightest members. Class O stars are comparatively rare, with the Milky Way Galaxy being thought to contain only some 20,000 of them. Contrast that with the (at least) 200 billion stars that the Milky Way Galaxy contains in total. In other words, only about one in ten thousand stars in the galaxy are Class O stars, and yet here in NGC 6231 you see lots of them!

NGC 6242

This open cluster in Scorpio is another southern sky object mid to far northern observers won't be able to see. That's a shame, because under the right conditions it is a very pretty object indeed, as well as an interesting one.

NGC 6242 is about 3,600 light years away and 9.5 light years in diameter. It is of middling age, around 50 million years old, and because of that, its brightest stars have long since burned themselves out (or more precisely, become supernovae). The result is a cluster that isn't particular dense or striking, with some 23 known members, and it isn't easy to tell the stars that make up the cluster from the stars in the background. That's a problem with all open clusters, but because NGC 6242 lies on the galactic plane, there are a whole lot more background stars to deal with!

There are also clouds of dust on the galactic plane as well, and NGC 6242 actually appears slightly dimmer than it should be precisely because of this. Astronomers know how bright the stars inside the cluster are, but they appear a little bit dimmer, and that's because there's dust obscuring that light.

NGC 6254 (M10)

M10 is a globular cluster in Ophiuchus only 3° from M12 (see the entry for NGC 6218). It is an easy object for telescopes of all sizes, and even binoculars will show the object as a small blob of light. Small to medium aperture telescopes should show its core is substantially brighter than the rest of the globular cluster, but apertures upwards of 150 mm are needed to resolve a good proportion of its constituent stars. With averted vision, a 200-mm SCT will resolve M10 almost down to the core.

This globular cluster is about 14,400 light years away and measures about 80 light years in diameter. At the eyepiece it is a fairly large object with an angular diameter of about 20 arcminutes, though this is much less than the 36 arcminutes diameter of the Great Globular Cluster in Hercules, M13. It is rated as a Class VII object on the Shapley–Sawyer scale.

NGC 6266 (M62)

M62 in Ophiuchus is one of the more peculiar globular cluster in the Messier catalog of deep sky objects. Whereas most globular clusters are spherical or elliptical in shape, this one is decidedly irregular, with its most concentrated region distinctly off to one side of the center. The likely explanation here is that M62 has been distorted by the tidal forces resulting from the gravitational pull of the galactic core. All globular clusters orbiting the Milky Way experience some degree of distortion because of this, but M62 is much closer to the core than most of the others, a mere 6,100 light years from the core, and consequently it experiences a much higher degree of deformation.

Mid to far northern observers will find this object challenging because it never rises very far about the horizon. In Southern England for example, it barely attains an altitude of 8° above the horizon.

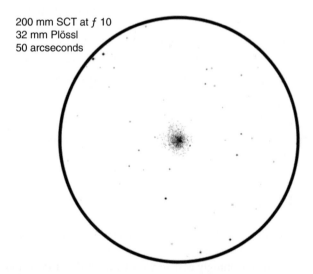

200 mm SCT at ƒ 10
32 mm Plössl
50 arcseconds

Figure 4.3. M62 is distinctive in lacking the near-spherical shape typical of most globular clusters. Unfortunately, it's southerly location makes it a difficult target for mid to far northern observers. (Image produced using Starry Night Pro. AllSky data courtesy of Main-Sequence Software Inc.)

M62 is about 22,500 light years from Earth and has a diameter of about 100 light years. It is a compact globular cluster, a Class IV object on the Shapley–Sawyer scale.

NGC 6273 (M19)

One of numerous globular clusters in Ophiuchus easily visible with a small telescope, M19 is most remarkable for its highly elliptical shape. This is presumably because M19 is a mere 5,200 light years from the galactic core, so that tidal forces are able to deform its shape. While close to the core, it's a fair distance from Earth, about 28,000 light years away, so this globular clusters is one of the fainter ones on the Messier list.

On the Shapley–Sawyer scale M19 is rated as a Class VIII object. This would suggest that it is fairly easy to resolve, but in practice this doesn't tend to be the case. While certainly visible through small telescopes where its granular texture will be apparent, you need a reasonably large aperture to begin to resolve this cluster, upwards of 150 mm under good conditions, and likely rather more under suburban skies.

For mid to far northern, observers, M19 can be a tough object. Observers in southern England will find it only gets to about 12° above the horizon.

NGC 6281

NGC 6281 is an open cluster in Scorpius far south of the celestial equator. It contains about 70 stars spread out across 20 arcminutes. It is quite large and bright, but because of its southerly location, not well known. Needless to say, the big problem for northern hemisphere observers is that this southern sky object doesn't rise very above the horizon. Even as far south as central Florida it only gets to about 25° above the horizon.

NGC 6281 is about 1,800 light years from Earth and is believed to be a little over 4 light years in diameter. It is about 220 million years old. Unusually for an open star cluster, it seems to contain at least one star that is an X-ray source. Long-exposure photographs reveal that there is some nebulosity associated with the cluster.

NGC 6341 (M92)

M92 is one of the most overlooked globular clusters in the sky, and a particular favorite of the author's. It is conveniently placed high up in the sky through much of the year, making it refreshingly easy to view compared to those globular clusters located in the more southerly constellations.

This is quite a bright globular cluster, and easily detected with binoculars; some astronomers even maintain they can see it with the naked eye (though that's unlikely to be the case under suburban skies). Through even a small telescope its granular texture is obvious, and the bigger the telescope, the better the view. Averted vision helps too, allowing more stars to be glimpsed, resulting in an altogether better impression of this cluster.

Since this globular cluster is in the constellation of Hercules, it is inevitably compared with M13, better known as the Great Globular Cluster. The two globular clusters are of similar age, each about 14 billion years old, though the precise value, as is often the case with globular clusters, is a matter or some debate. In terms of absolute size, they aren't very different either, M13 measuring about 140 light years in diameter and M92 only very slightly less, at about 110 light years.

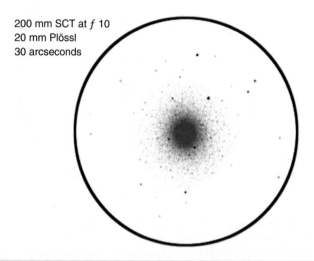

200 mm SCT at *f* 10
20 mm Plössl
30 arcseconds

Figure 4.4. M92 is a superb globular cluster, but underrated because of its proximity to the Great Globular Cluster in Hercules. (Image produced using Starry Night Pro. AllSky data courtesy of Main-Sequence Software Inc.)

So why is M13 so much brighter and more impressive? One reason is that M13 contains rather more stars than M92, with the former having about 400,000 stars to its name and M92 nearer 300,000. M13 is also a bit closer, at only 25,000 light years from Earth, versus almost 26,700 light years in the case of M92. Finally, M13 is a bit less concentrated, being accounted a Class V on the Shapley–Swayer scale compared with M92, which rates as a Class IV object. So while the two objects aren't too dissimilar on paper, in practice one definitely outshines the other.

But all that said, M92 remains an extremely attractive globular cluster. It also has a very interesting claim to fame. As Earth wobbles on its rotational axis, a process called precession, M92 is periodically brought within a degree or so of the celestial pole; in other words, it occupies the same position as the Pole Star (Polaris) does today. Unfortunately, this isn't something readers of this book will have seen before or are likely to see in the future: the last time it happened was about 11,500 years ago, and it'll be another 14,300 years before it happens again!

NGC 6514 (M20, Trifid Nebula)

This nebula in Sagittarius has both an emission component (which is easy to see) and a reflection component (which is not).

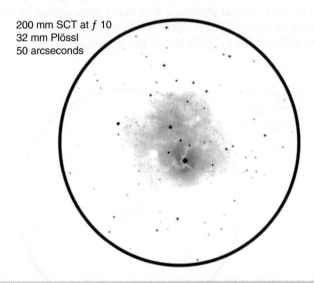

200 mm SCT at ƒ 10
32 mm Plössl
50 arcseconds

Figure 4.5. For observers at mid to low northern latitudes, the spectacular Trifid Nebula is one of the highlights of the year, almost the equal of the Orion Nebula in terms of size and interest. (Image produced using Starry Night Pro. AllSky data courtesy of Main-Sequence Software Inc.)

The emission nebula at least is a good target for telescopes of all sizes, and while comparatively small, about 11 arcminutes across, but it is bright and responds extremely well to the use of a narrowband or O-III filter. An H-beta filter can also be worthwhile when observing the Trifid Nebula. Like the Orion Nebula, this is a nebula that looks good almost regardless of aperture. Through small aperture, wide field telescopes it's an attractive sight when set against the background stars, while higher magnifications and larger apertures reveal more of the detail, including the dust lanes that give the Trifid Nebula its distinctive three-lobed appearance.

The reflection nebula is rather more difficult to see. It is set to one side of the emission nebula, centered around a single bright star, SAO 186149. Whereas the emission nebula is emitting its own light because it contains ionized gas, the reflection nebula is merely reflecting starlight. As such, this object responds poorly to the use of narrowband or O-III filters. Indeed, you may find yourself shuffling between the use of a filter to see the emission nebula better, and then taking the filter away so you can try to discern the reflection nebula.

Astronomers have studied the Trifid Nebula intensively, and it is one of the objects imaged by the Hubble Space Telescope. The Trifid Nebula is revealed as a stellar nursery, filled with regions where stars are being formed from clouds of gas and dust.

The Hubble photographs show peculiar structures known as stellar jets, streams of glowing gas being pushed out of the main body of the nebula by a forming star. It's believed that as stars develop, they emit vast amounts of ultraviolet light. This heats up the gas around the young star, and the hot gas bubbles outwards from the surface of the nebula like steam bubbling out of a pool of boiling mud. When this happens a pock mark of sorts is left behind, and more gas and dust streams out of the hole, glowing as it does so because of the energy it receives from the developing star; these streams are stellar jets. But because these jets allow material to leave the cloud, they effectively limit how big the star inside the nebula can get by depriving it of the material it needs to grow.

What's interesting about this is that it explains why not all stars become giants, and how some, deprived of the material they need to become giants, end up as dwarf stars like our own Sun.

It's a shame that far northern observers will find the Trifid Nebula a difficult target. In Southern England it only gets to about 15° above the horizon, making a clear horizon free of light pollution crucial. Observers farther south will find things a lot easier though, and excellent views can be had even from the Midwest of the United States.

The distance between Earth and the Trifid Nebula is not clearly known, but is thought to be around 5,000 light years, though estimates vary from as little as 2,200 light years up to as much as 9,000 light years. While its angular diameter of about 28 arcminutes is easy enough to measure, this can't easily be turned into an absolute size without knowing how far away it is. Obviously the farther away it is, the bigger it would have to be to have the same angular diameter. Many astronomers think that it is about 25 light years across, but that number may well be revised up or down once its true distance can be established.

NGC 6523 (M8, Lagoon Nebula)

This spectacular emission nebula in Sagittarius is large and comparatively nearby, estimated at a distance of some 5,200 light years from Earth and thought to be about 140 light years in width. The Lagoon Nebula is an object that responds incredibly well to the use of narrowband and O-III filters, both types of filter greatly enhancing its contrast and making detail much easier to see. On the other hand, filters dim stars somewhat, so you'll want to switch between using a filter and not using a filter depending on whether you're concentrating on the nebula or the stars within it.

Like other emission nebulae, the Lagoon Nebula is a huge cloud of ionized gas. The two brightest stars that are visible within it are providing the energy that causes the gas to glow. Long-exposure photographs also reveal the presence of structures called Bok globules, opaque clouds of gas and dust that block light, and so appear as black specks set against the brightly glowing nebula. These are thought to be clouds of gas and dust that are collapsing in on themselves, and perhaps taking their first steps towards becoming a star.

The open cluster NGC 6530 can be easily seen to one side of the Lagoon Nebula. It is an extremely young cluster of stars, condensing out from the Lagoon Nebula only about 2 million years ago.

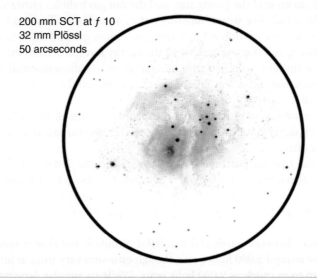

200 mm SCT at *f* 10
32 mm Plössl
50 arcseconds

Figure 4.6. M8 is a large emission nebula in Sagittarius that can be difficult to see from suburban locations without the use of a narrowband or O-III filter. (Image produced using Starry Night Pro. AllSky data courtesy of Main-Sequence Software Inc.)

NGC 6541

This bright globular cluster in Corona Australis is unfortunately not visible from mid to far northern locations, and even as far south as central Florida it only rises about 19° above the horizon. So while an easy object for binoculars and a lovely sight through telescopes of all sizes, NGC 6541 isn't a well known object and consequently often overlooked.

NGC 6541 has an angular diameter of about 15 arcminutes, so it's pretty small compared to some of the other showpiece globular clusters of the season. But it is quite bright and has a very condensed core, so it is easy to see even under somewhat light-polluted conditions. NGC 6541 is a Class III object on the Shapley–Sawyer scale, and resolving all but the outer whorls of stars is difficult.

NGC 6541 is thought to be about 22,800 light years away from Earth. Its age is a matter of some debate because this cluster hasn't been well studied, but one study in 1997 suggested an age of between 16 and 17 billion years. This is quite a bit older than current estimates for the age of the universe, which is thought to be between

13.5 and 14 billion years old. Needless to say, globular clusters aren't older than the universe, and the fact that they seem older says more about the application of various assumptions and constants to the calculations used to estimate their age. Whatever the precise value turns out to be, it's fairly certain that NGC 6541 will be a very ancient object only a little younger than the universe, as is the case with most other globular clusters.

NGC 6603

NGC 6603 is an open cluster within the Sagittarius Star Cloud (sometimes referred to as M24). Whether or not NGC 6603 is easy to spot depends largely on how good your view of the Sagittarius Star Cloud happens to be. From far northern locations views of objects in Sagittarius tend to be obscured by light pollution and dust in the part of the sky just above the horizon where this constellation appears. In such places, picking out NGC 6603 isn't easy. But more southerly locations enjoy better views of the Sagittarius Star Cloud, in which case NGC 6603 will be seen as definite bright patch within the rich field of stars that makes up this particular star cloud.

The cluster itself contains at least 30 stars and has a diameter of about 14 light years. It is of middling age, around 100 million years old, and thought to be about 12,000 light years away.

The Sagittarius Star Cloud looks like a cloud of stars from our vantage point on Earth, but it is in fact the Sagittarius-Carina Arm of the Milky Way Galaxy. Our planet Earth orbits a star situated on a short, spur-like arm of the Milky Way known as the Local Spur. On either side of this spur are two of the Milky Way Galaxy's four major arms, the Perseus Arm in the direction of intergalactic space, and the Sagittarius-Carina Arm in the direction of the galactic core. That we see this arm in chunks like the Sagittarius Star Cloud rather than a single continuous stream of stars is down to the presence of clouds of dust in between our Local Spur and the Sagittarius-Carina Arm.

Besides NGC 6603, other objects in the Sagittarius-Carina Arm include the Trifid Nebula (NGC 6514), the Wild Duck Cluster (NGC 6705), and the Lagoon Nebula (NGC 6523).

NGC 6618 (M17, Horseshoe Nebula, Lobster Nebula, Omega Nebula, Swan Nebula)

This interesting object has been known by several names over the years. But to quote the immortal bard, *What's in a name? That which we call a rose by any other name would smell as sweet.* This is indeed one of the loveliest deep sky objects, a combination of an open star cluster with an emission nebula similar in type to objects like the Lagoon Nebula and the Orion Nebula, but with a charm all its own.

M17 looks good through telescopes of all sizes, appearing as a small L-shaped nebula with a scattering of bright stars set within it. Dark skies are helpful when viewing this object, since without dark skies spotting the nebula will be difficult. Under suburban conditions the use of some type of light pollution filter may be necessary; either a narrowband filter or an O-III might be used, but the flip side to such filters is that the open star cluster will be dimmed and your overall impression of the object somewhat lessened. On the other hand, filters make it easier to see the full extension of the nebula rather than just the brightest bits.

The nebula is a site of active star formation, including the cluster of about 35 bright stars we can see embedded in the nebula. The nebula seems to be bigger and more massive than the Orion Nebula, but being farther away, at least 5,000 light years versus around 1,500 light years for the Orion Nebula, it appears smaller and less brilliant. The bright part of the M17 nebula that we can see is probably about 15 light years in width, but the whole thing is much bigger and more like 40 light years in width.

Like most objects in the Sagittarius-Carina Arm, getting good views of M17 can be tricky for far northern observers. A good clear horizon is important. But it does at least rise reasonably high in the sky; from locations in southern England for example, M17 may reach as much as 22° above the horizon, which should be enough to ensure tolerably good views.

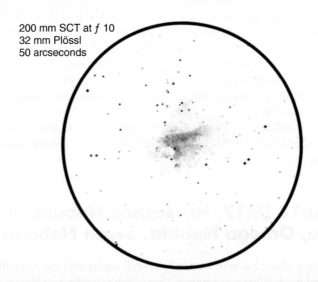

200 mm SCT at ƒ 10
32 mm Plössl
50 arcseconds

Figure 4.7. M17 includes an open star cluster and an emission nebula. A narrowband or O-III filter will make the nebula easier to see under suburban conditions. (Image produced using Starry Night Pro. AllSky data courtesy of Main-Sequence Software Inc.)

NGC 6626 (M28)

M28 is an attractive globular cluster in Sagittarius that might not be quite so impressive as M22 (NGC 6656) but is a very nice deep sky object for medium the large aperture telescopes nonetheless. It is relatively small though, with an angular diameter of about 11 arcminutes, so reasonably high magnifications are essential if you want to begin resolving the cluster into individual stars.

M28 is about 18,300 light years away and has a diameter of about 60 light years. It is rated as a Class IV globular cluster on the Shapley–Sawyer scale, putting it towards the more condensed end of the spectrum. Even with a big telescope this is a difficult object to fully resolve.

Within this globular cluster lies a millisecond pulsar, a pulsar that appears to flash its beam of radio waves at as us hundreds of times per second (regular pulsars flash somewhat less frequently, up to once every 8 s or so). Pulsars are rapidly rotating neutron stars, neutron stars being the remnants of a massive stars that collapsed in on themselves after they became supernovae at the end of their lives. Neutron stars are small, typically only about 20 km in diameter. But they are also incredibly massive, the average neutron star being substantially more massive than the Sun. One famous way to describe this colossal density is to observe than a single teaspoon of the stuff within a neutron star would weigh a billion metric tons!

What makes a pulsar different for a run-of-the-mill neutron star is that it emits an extremely regular pulse of radio waves that we can detect on Earth using radio telescopes. All neutron stars rotate rapidly, typically several times per second, though as they age, they slow down, and the slowest ones seem to take about 8 s or so to rotate. Most (though seemingly not all) neutron stars emit radio waves, but pulsars are different because their axis of rotation isn't aligned with their magnetic field. The magnetic field acts like a sort of funnel, directing radio waves outwards from the magnetic north and south poles, but because the pulsar is rotating about a different axis, its axis of rotation, the north or south pole only points towards Earth at particular intervals. Because pulsars are very dense their rotational speed is extremely regular, and the result is what has been called a stellar lighthouse, or more precisely, a very regular flash of radio beams.

Millisecond pulsars, like the one detected in M28, are rotating so incredibly quickly that they flash hundreds of times per second. How they evolve is a bit of a mystery (indeed, much about how pulsars are formed and work remains obscure). Since globular clusters like M28 are known to be very ancient, it was assumed that any pulsars they once contained would have faded away billions of years ago. Perhaps what happens is that the closeness of individual stars in globular clusters allows pulsars to accrete material from nearby stars, and so prolong their lives. M28 was the first globular cluster within which a millisecond pulsar was detected, though others have been now detected, both in this globular cluster and in others.

Far northern observers will find this globular cluster a difficult target. From southern England it doesn't get to more than about 15° above the horizon. Observers farther south will of course find it a much easier and more rewarding target.

NGC 6656 (M22)

This large, bright globular cluster in Sagittarius is one of the best in the sky, but its southerly location makes it a difficult target for far northern observers. So while some would argue it is a better globular cluster than the Great Globular Cluster in Hercules M13, it's that latter cluster that tends to be considered as the best globular cluster for northern hemisphere observers.

In any case, if you can get a good view of this globular cluster, then you're going to be in for a treat. It is extremely bright (magnitude 5.1) and looks good through even relatively small telescopes. Medium to large aperture instruments, from 150 mm apertures upwards, will reveal a large, bright blur of light that can be relatively easily resolved into individual stars, particularly with averted vision. On the Shapley–Sawyer scale M22 is rated as a Class VII object.

M22 has an angular diameter of 32 arcminutes, similar to that of the Full Moon, but it is of course much farther away, and consequently very much larger in absolute terms. Current estimates suggest a distance of a little over 10,000 light years and a diameter of almost 100 light years. M22 contains about half a million stars. Surprisingly perhaps given its brightness, it is actually partially obscured by a cloud of dust in between us and the globular cluster.

M22 is unusual among globular clusters in containing a planetary nebula. It also happens to lie less than a degree away from the ecliptic, so the Moon and planets often pass close by.

NGC 6705 (M11, Wild Duck Cluster)

The Wild Duck Cluster in Scutum is one of the richest open clusters easily visible with small telescopes, with almost 3,000 constituent stars. It is also remarkably compact, with an angular diameter of just 13 arcminutes, and the Wild Duck Cluster can be easily confused with a globular cluster at first glance.

The Wild Duck Cluster is at least 5,500 light years away and more than 15 light years in diameter, though different astronomers have different ideas about precisely how big and how far away it is. It contains mostly hot blue stars it has some orange and red giant stars as well. It is the presence of these older stars that lead astronomers to believe that it is of fairly advanced age by open cluster standards, around 250 million years old. In photographs the variety of colors are easy to see, but this is less obvious when viewed through small to medium aperture telescopes, particularly under suburban sky conditions.

Its common name comes from the wedge-shaped arrangement of the stars that some have compared to a flock of ducks. The stars look remarkably close together, and indeed they are. On average, the stars in this cluster will only be about a light year apart, a trivial distance compared with the usual distances between stars throughout the rest of the galaxy.

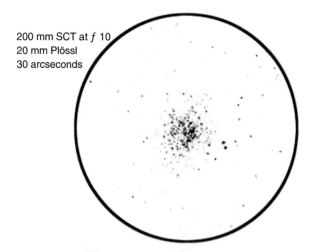

200 mm SCT at *f* 10
20 mm Plössl
30 arcseconds

Figure 4.8. The Wild Duck Cluster is a very rich open cluster that might easily be mistaken for a globular cluster at first glance. (Image produced using Starry Night Pro. AllSky data courtesy of Main-Sequence Software Inc.)

NGC 6720 (M57, Ring Nebula)

The Ring Nebula in Lyra is probably the most enjoyed planetary nebula among amateur astronomers. Although small, it is very bright, and conveniently placed high in the sky for a large part of the year. Like most planetary nebulae, this is an object that responds very well to the use of a narrowband or O-III filter, but the Ring Nebula is sufficiently bright that under reasonably good suburban skies it is easy to see without a filter.

As its name suggests, the Ring Nebula has an elliptical shape. Very small telescopes may not reveal this shape easily, but anything upwards of 90 mm should do so without problems. Because the Ring Nebula is very small, barely an arcminute in diameter, moderate to high magnifications are essential. You van certainly see the object with lower magnifications, but picking it out from the stars in the same patch of sky will be difficult. Use of a narrowband or O-III filter will of course remove any doubt, dimming the background stars greatly but hardly affecting the nebula at all.

Given sufficient magnification the shape of the elliptical ring of ionized gas can be easily seen. The star at the center of the ring is much more difficult to spot though. At magnitude 14.7 it is essentially out of the range of most commercial go-to telescopes, and is typically seen in telescopes with apertures substantially above 300 mm.

The Ring Nebula is about 2,300 light years away and the bright part of the nebula that we can see is a little less than a light year in diameter.

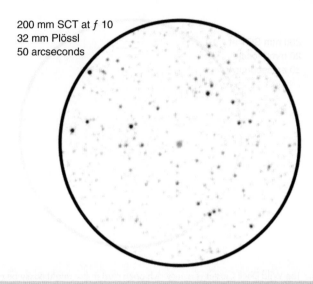

200 mm SCT at f 10
32 mm Plössl
50 arcseconds

Figure 4.9. The Ring Nebula is a very small object and reveals its distinct shape best under high magnification. (Image produced using Starry Night Pro. AllSky data courtesy of Main-Sequence Software Inc.)

NGC 6838 (M71)

This globular star cluster in Sagitta is not often observed, but it's a remarkably nice little globular cluster easily seen as a moderately bright hazy patch of light through small telescopes. Unlike most other globular clusters, there isn't an obvious division between a circular core that is too condensed to resolve and a more easily resolved halo of stars around it. Instead M71 is somewhat irregular in shape and doesn't have an obvious core at all. Indeed, through a small telescope it might almost be mistaken for a planetary nebula!

In fact the nature of this cluster has been much discussed over the years, with some astronomers arguing that it is a very loose globular cluster, while others describe it as a very condensed open cluster. As such, it makes an interesting contrast with the Wild Duck Cluster, NGC 6705, another cluster that seems to straddle the boundary between these two classes of deep sky object.

One reason it appears relatively loose by globular cluster standards is that it is very close to Earth, at a little over 12,700 light years away. Despite being so nearby, its angular diameter of about 7 arcminutes is rather low; compare this with the Great Globular Cluster in Hercules that is about twice as far away but still measures a whopping 36 arcminutes in diameter. So by globular cluster standards M71 is apparently very small, only 27 light years in diameter (though some astronomers believe that there are fainter stars associated with the cluster much farther out, and it may be as much as 90 light years in diameter if these outliers are included).

Another difference between M71 and most other globular clusters is in its age, a mere 9–10 billion years. Most globular clusters are a fair bit older, though this is still very much older than most open clusters. Curiously, M71 isn't often given a rating on the Shapley–Sawyer scale, an indication of the uncertainty that has surrounded the nature of this strange little cluster over the years. On the occasions when it has been rated, it's been considered a Class X or Class XI globular cluster.

NGC 6853 (M27, Dumbbell Nebula)

Second only to the Ring Nebula (NGC 6720) among the summer season planetary nebulae, this bright and surprisingly large object in Vulpecula is an easy target for telescopes of all sizes. Its common name refers to the shape of the two brightest portions, which do indeed form a dumbbell, or at least bow-tie, shaped structure. Long-exposure photographs reveal that the nebula is actually rather more elliptical than it seems visually.

The Dumbbell Nebula responds very well to the use of narrowband and O-III filters. Such filters will be extremely useful when this object is viewed under suburban conditions. Filters make it easier to see the fainter parts of the nebula.

Large aperture telescopes can reveal the magnitude 13.5 star at the center of the nebula, but this typically needs dark skies to be possible with telescopes around the 200 mm mark. The bigger the telescope, the easier it is to see this star. This star is tiny, having a radius of about 38,000 km. This is one eighteenth the size of the Sun, which has a radius of about 695,500 km. In fact it is more planet-sized than star-sized, with a radius about halfway between that of Saturn (60,200 km) and Uranus (25,500 km).

The distance between Earth and the Dumbbell Nebula is somewhat uncertain. Its brightness and large apparent size imply that it is relatively nearby, but estimates of its distance varies from less than 1,000 light years to over 4,000 light years. Because we don't know how far away it is, it's difficult to translate its angular diameter into an actual length. But if it is about 2,300 light years away, as many astronomers believe, then the diameter of the brightest part of the nebula, the bit we can see with our telescopes, is about 0.9 light years.

Because the angular diameter of the Dumbbell Nebula has been recorded by astronomers over many years, it has become clear that it is expanding quite quickly. The expansion rate is about an arcsecond per century, and from that astronomers have been able to calculate how long this nebula has been in existence, assuming a uniform expansion rate. Using this method, the Dumbbell Nebula is believed to be no more than 14,600 years old.

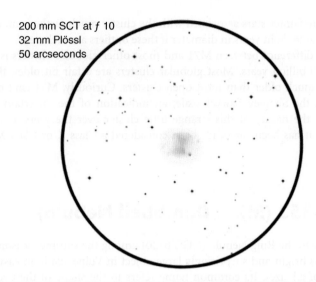

200 mm SCT at ƒ 10
32 mm Plössl
50 arcseconds

Figure 4.10. The Dumbbell Nebula is relatively larger and bright, but a narrowband or O-III filter will make the object easier to see and more detailed. (Image produced using Starry Night Pro. AllSky data courtesy of Main-Sequence Software Inc.)

NGC 6960, NGC 6979, NGC 6992 and NGC 6995 (Veil Nebula, Cygnus Loop, Network Nebula)

The Veil Nebula in Cygnus is a supernova remnant and a wonderful target for telescopes of all sizes, but it is one of those objects where using the right filter makes all the difference. Even under quite dark skies the Veil can be difficult to see without a filter, but with an O-III filter it can, under reasonably dark skies, become an easy and rewarding target even with a small telescope. Some of the best views the author has ever had of this object were under exurban skies (3–4 on the Bortle Dark Sky scale) using nothing more than a 76 mm-wide field refractor, a 35 mm-Panoptic, and an O-III filter. On the other hand, under more light polluted skies (around a 6 on the Bortle scale) that same combination barely reveals the Veil at all, and even when the filter is used with an aperture of 200 mm, only the brightest parts of the Veil are glimpsed.

Four NGC objects are included in the Veil, though only two of them are (comparatively) easy targets, the other two being much fainter and more difficult to see except under good, dark skies. The brightest part of the Veil is the Eastern Veil, consisting of NGC 6992 and NGC 6995. Slightly less bright is the Western Veil, NGC 6960. Taken together, the Eastern and Western Veil seem to make distinct arcs along an otherwise invisible ellipse some 230 by 160 arcminutes in size, hence the alternative name for this nebula, the Cygnus Loop. The final component is NGC 6979, sometimes called Pickering's Wisp, a notoriously difficult portion to see visually, and really only a viable target for astronomers with large telescopes observing under very dark skies.

The bit to aim for first is the Eastern Veil, either NGC 6992 or NGC 6995. Since this structure is more than a degree in length, you want to use the widest possible field of view. On a 200 mm-SCT, an f/6.3 reducer-corrector with wide-field eyepiece like a 35 mm-Panoptic offers a field of view of about 2°, which should be enough to bring the Eastern Veil into view. Even with an O-III filter, the contrast between the nebula and the background sky may not be very great if light pollution is at anything more than trivial levels. Try using averted vision, or else remove the filter and move it in and out of the space between the eyepiece and your eye. With luck, the Veil will flicker into view.

If the Eastern Veil is visible, then try for the Western Veil! It's similar in size to the Eastern Veil, but its arc shape curves in the opposite direction. Unlike the Eastern Veil, which is certainly visible from suburban locations, the Western Veil is a tough object under suburban skies, even with an O-III filter.

The Veil Nebula was formed by a supernova some 10,000 years ago. At the time, the supernova must have been incredibly bright, but it has long since faded away to obscurity. The Veil is about 2,500 light years away and something like 100 light years across.

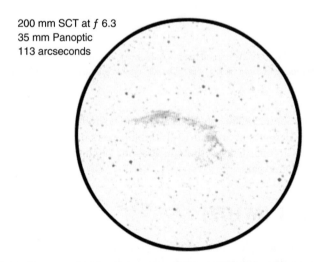

200 mm SCT at *f* 6.3
35 mm Panoptic
113 arcseconds

Figure 4.11. When an O-III filter is used, the Eastern Veil Nebula NGC 6992 goes from being an impossible target to merely a very difficult one. It is surprisingly large though, and a wide field eyepiece is absolutely essential. (Image produced using Starry Night Pro. AllSky data courtesy of Main-Sequence Software Inc.)

Interesting Deep Sky Objects

NGC 5986

This globular cluster in Lupus is a bit too far south for observers at mid to far northern locations to enjoy, but at lower latitudes it is well worth adding to your observing list. It's a small but comparatively bright object a little under 10 arcminutes in diameter. Medium aperture telescopes clearly reveal its granular texture and its dense core.

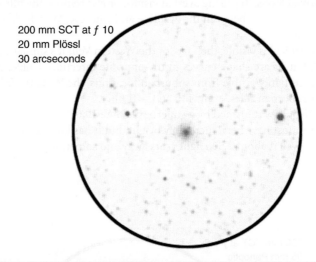

200 mm SCT at *f* 10
20 mm Plössl
30 arcseconds

Figure 4.12. NGC 5986 is a small but bright globular cluster in the southerly constellation of Lupus. (Image produced using Starry Night Pro. AllSky data courtesy of Main-Sequence Software Inc.)

NGC 5986 is about 34,000 light years away, and is similar in age to the Great Globular Cluster in Hercules at around 13–14 billion years old. It is quite a bit smaller than the Great Globular Cluster though, at about 100 light years in diameter compared with 160 light years.

NGC 6171 (M107)

The constellation of Ophiuchus contains several globular clusters, some of which make it into the first rank while others, like M107, are of the second rank when it comes to impressiveness. Compared with most of the other globular clusters in the Messier Catalogue, this one is fairly dim and difficult to see. Under good dark skies it can be spotted with a small telescope, but under suburban conditions it is a challenge for anything with an aperture of less than 150 mm. Even with a 200 mm telescope, suburban observers are going to find resolving all but the outermost whorls of stars difficult.

M107 is fairly close to Earth at a mere 21,000 light years distant, but because of clouds of dust between us and the cluster, it's not nearly as bright as it should be. It is a fairly loose globular cluster, rated as a Class X object on the Shapley–Sawyer scale, and with a large telescope is fairly easy to resolve. Its apparent size is very great though, a mere 13 arcminutes, so substantial amounts of magnification will be required. In absolute terms its of middling size, at something like 80 light years in diameter.

NGC 6333 (M9)

This globular cluster in the constellation of Ophiuchus only has an angular diameter of 12 arcminutes, but it is bright and usually quite easy to see. It does have a relatively southerly position though, and from far northern locations it may be difficult to see through the haze and light pollution along the horizon. In southern England, for example, it only gets to about 20° above the horizon, so while a perfectly viable target for observers in rural or perhaps exurban environments, suburban astronomers will find it a much more elusive target.

M9 is one of the closest globular clusters to the galactic core, only 5,500 miles out from the center of the galaxy. It is about 26,000 light years from Earth, and also happens to be lying behind some clouds of interstellar dust, so its overall brightness is a bit less than would otherwise be the case for a globular cluster of this size and proximity. While it can be glimpsed through small to medium aperture telescopes without too much difficulty, a large aperture telescope is really needed to fully appreciate this globular cluster. M9 is rated as a Class VIII object on the Shapley–Sawyer scale, and a 200 mm or larger telescope will reveal a compact core surrounded by a much looser halo of stars around the outside.

M9 is about 90 light years in diameter and has an age of about 14 billion years.

NGC 6402 (M14)

This slightly elliptical globular cluster in Ophiuchus can be a tricky object to see, and has a reputation for being rather less rewarding than many of the other globular clusters in this constellation.

The problem isn't so much its size, which at about 11 arcminutes in diameter is comparable to many of the other second-rank globular clusters, but its brightness, which is a full magnitude less than that of the showpiece globular clusters in Ophiuchus like M19 (NGC 6273). In practice this globular cluster is visible through even quite small telescopes, but doesn't start to become resolved until apertures of at least 200 mm are used.

M14 is a little over 30,000 light years away and believed to have an actual diameter of about 100 light years. It is among the few globular clusters within which a nova has been observed. Although the nova was photographed in 1938, it wasn't noticed until the photographic plates were examined in 1964. M14 is a Class VIII object on the Shapley–Sawyer scale.

NGC 6405 (M6, Butterfly Cluster)

M6 is an open cluster in Scorpius consisting of some 80 stars across a region a little over 50 arcminutes in diameter. Although not especially impressive, it is an attractive sight through telescopes of all apertures, and the presence of a bright orange supergiant BM Scorpii, makes it noteworthy. This star is a long period variable star, fluctuating between magnitudes 5.5 and 7 over a period of 850 days (about 28 months).

There is some debate over its size and distance from Earth, though the most recent estimates suggest a distance of about 1,600 light years and a diameter of about 12 light years. M6 is believed to be about 100 million years old, though again, there's a degree of uncertainty over the precise value.

Both M6 and the nearby open cluster M7 (NGC 6475) are naked eye objects under dark sky conditions. They were known at least as far back as the second century A. D., when the Greek astronomer Ptolemy recorded them in his influential survey of the heavens, known today by its Arabic name, the *Almagest*. M6 is often called the Butterfly Cluster on account of its shape, but M7 is known as Ptolemy's Cluster in reference to this early report of its existence.

M6 is an extremely difficult target for far northern observers. Although it may rise above the horizon, it doesn't rise very high; from southern England, it attains an altitude of about 3.7° above the horizon.

NGC 6445 (Box Nebula)

This a small planetary nebula in Sagittarius is normally considered a target for astrophotography rather than visual observation, but it is fairly bright and quite easy to spot given the right conditions. The trick is telling it apart from a star, because the nebula is very small and looks like a fuzzy star at low to medium magnifications. The best approach is to contrast it with the magnitude 8.8 star SAO 160841 just over 10 arcminutes away. Try using a narrowband or O-III filter if NGC 6445 isn't obvious; with the filter in use, the star will become much dimmer, while the planetary nebula will hardly be affected at all.

Even with high magnifications, there isn't much to see though; NGC 6445 is barely 3 by 0.9 arcseconds in size and as such is too small for anything beyond its general shape to be obvious.

NGC 6445 is a tough but not impossible target for observers at far northern locations, getting to about 18° above the horizon from the point of view of an observer in Southern England. Observers farther south will have a much easier time of things.

It is generally difficult to estimate the distance between Earth and planetary nebulae, and NGC 6445 is no exception. So while thought to be about 4,500 light years away, this number is likely to be revised over time.

NGC 6475 (M7)

This open cluster covers about 1.3° of the sky and is a particularly nice object for telescopes equipped with wide-field eyepieces. An f/10 200-mm SCT won't be able to frame the whole thing in the field of view of a 1.25-in. eyepiece, but a combination including an f/6.3 reducer-corrector and an eyepiece that produces a 2° field of view, such as a 35-mm Panoptic will set M7 against a beautifully rich field of stars.

Because the cluster contains eight stars including many that are comparatively bright, it's quite an easy cluster to pick out and looks good through telescopes of all apertures, including binoculars. Unfortunately for far northern observers, this cluster is too far south to be easily seen from such latitudes. From southern England, for example, it rises little more than 3° above the southern horizon.

M7 is a large and bright object because it is not far from Earth, only about 800–1,000 light years in fact. Depending on how far away it is, and there's some debate over the precise value, it probably measures about 20–25 light years in diameter. M7 is a middling age open star cluster, and thought to be about 220 million years old.

Like M6 (NGC 6405), M7 is visible to the naked eye under good conditions as a faint patch of light. It has been known since ancient times, and was included by Ptolemy in the *Almagest*. For this reason, it is sometimes called Ptolemy's Cluster.

NGC 6494 (M23)

M23 is an impressive open cluster in the constellation of Sagittarius. It is a large object with an angular diameter of about 27 arcminutes, making it appear about the same size as a full Moon. That being the case, a wide field eyepiece is a good idea when observing this object, though failing that, a long focal length eyepiece such as a 32-mm Plössl will do the trick nicely.

Like many objects in Sagittarius, M23 is a difficult target for far northern observers. From localities in southern England, for example, it only gets to about 19° or so above the horizon. Although certainly not an impossible target, a clear horizon will obviously be essential, and ideally one with little or no light pollution.

M23 is about 2,100 light years away and contains about 150 stars spread out across some 15 light years of space. One of its quirks is a neat arc-shaped alignment of five stars close to its center.

NGC 6520

NGC 6520 is an open star cluster in Sagittarius. Although an attractive enough object in its own right, it is famous among amateur astronomers for containing one of the more easily observed dark nebulae. In this case, the dark nebula is known to amateurs as the Ink Spot, though its scientific designation is Barnard 86. Dark nebulae are clouds of dust that obscure the light from stars behind them.

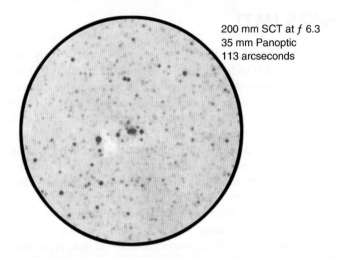

200 mm SCT at *f* 6.3
35 mm Panoptic
113 arcseconds

Figure 4.13. The open star cluster NGC 6520 (*at center*) is set within a very densely packed star field, but next to the cluster is a *dark nebula* known as the *Ink Spot* (the *bright white* patch to the *left* and *slightly below* NGC 6520 in this image). (Image produced using Starry Night Pro. AllSky data courtesy of Main-Sequence Software Inc.)

Far northern observers will find NGC 6520 a difficult target, since it only rises about 10° above the horizon from latitudes as far north as southern England. But from more favorable locations the open star cluster will be obvious as a collection of hot, young blue stars set against a dense field of older orange and yellow stars.

Whether or not you'll be able to see the Ink Spot will depend on how good your skies are. Under light-polluted suburban conditions, or when the object is low in the sky and viewed through dusty haze, the Ink Spot isn't at all easy to see. But if the conditions are right, the Ink Spot can be seen even through small telescopes as a black patch devoid of stars when compared to the dense star field that makes up the rest of the field of view.

NGC 6520 is about 5,500 light years away and 10 light years in diameter. The cluster contains about 60 members. As mentioned already, it is made up of young blue stars, and the yellow and orange stars that appear to be the same cluster are actually much older stars that just happen to lie in the same general direction.

NGC 6530

See NGC 6523, the Lagoon Nebula, in the showpiece objects section.

NGC 6544

NGC 6544 is a globular star cluster in Sagittarius. It is relatively small in terms of apparent size, a mere 4.6 arcminutes, but reasonably bright, and isn't too difficult to see given good conditions. Under suburban conditions the bright core at least should be visible through telescopes with apertures upwards of 150 mm.

This globular cluster is quite nearby, only about 8,800 light years away, but appears as a small object because it is indeed a relatively small globular cluster, only about 12 light years across. Compare this with the Great Globular Cluster in Hercules, M13 (NGC 6205), which is much farther away at about 25,000 light years distant, but being so much bigger, over 160 light years in diameter, it appears through the telescope as a far more imposing object.

NGC 6544 is a southern sky object, and is consequently difficult to see from far northern locations. From Southern England for example, it only gets to about 13.5° above the horizon. Observers farther south should not have much difficult though.

NGC 6572 (Blue Racquetball, Emerald Eye, Planet Krypton Nebula)

This planetary nebula in Ophiuchus is small but bright, and not difficult to see even under suburban conditions. Although NGC 6572 has an elliptical shape when viewed under steady conditions and high magnifications, because of its very small angular size, 7.8 arcseconds at its widest, it might easily be mistaken for an out-of-focus star. So as is generally the case with planetary nebulae, a good test is to use a narrowband or O-III filter, and see if the object dims. Stars will be dimmed significantly, but NGC 6572 will be hardly dimmed at all.

NGC 6572 is notable for being relatively colorful, appearing as either blue or green, depending on the observer, and has thereby acquired a variety of different names. More prosaically perhaps, it is one of those planetary nebulae that really does look similar to an actual planet, in this case resembling in apparent size and color the planet Uranus.

As is often the case with planetary nebulae, there's some uncertainty over how far NGC 6572 is from Earth, but current estimates suggest a distance between 4,000 and 5,500 light years. Interestingly, it is thought to be a remnant of a star very similar in mass to our own Sun, a star that died something like 2,600 years ago.

NGC 6633

NGC 6633 is a large, bright open cluster in the constellation of Ophiuchus highly regarded as a good target for binoculars and wide field refractors. It has an angular diameter of about 20 arcminutes, and consequently you want to use the lowest power eyepiece you have when looking at this object. On a 200-mm SCT, a 32-mm Plössl works very well. It looks even better if you can use an f/6.3 reducer-corrector to widen the field of view farther, and using a 2-in. eyepiece like a 35-mm Panoptic improves things even more.

NGC 6633 is a little over 1,000 light years away and contains at least 30 stars. Because it contains some orange giant stars as well as numerous blue stars, its age has been estimated at 660 million years, making it a middling age open star clusters. It's amusing to think about what was happening on Earth at this time. In terms of life, things were almost entirely limited to single-celled forms, and there were no true animals at this stage, though sponges of a sort existed. Some paleontologists also believe that Earth passed through the last of a series of Snowball Earth phases that saw the entire surface covered with ice.

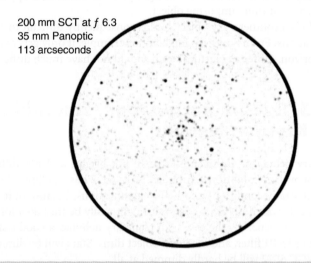

200 mm SCT at f 6.3
35 mm Panoptic
113 arcseconds

Figure 4.14. NGC 6633 is a nice open cluster best viewed through a low power, wide field eyepiece. (Image produced using Starry Night Pro. AllSky data courtesy of Main-Sequence Software Inc.)

NGC 6637 (M69)

M69 is a relatively dim globular cluster, and like many objects in Sagittarius, it is difficult to see from far northern locations. Observers in southern England will find that it only rises about 5° above the southern horizon, and for all practical purposes this is likely to make the object invisible.

M69 is quite small, only about 42 light years in diameter, and it is also fairly distant, 28–30 light years away from Earth, The end result is a small object only 8.3 arcminutes in diameter. Despite its size, it is bright enough be seen (if not resolved) through small telescopes as small as 100 mm, perhaps less should the sky be sufficiently dark.

Telescopes around 200 mm reveal the spherical shape and allow the outer whorls of stars to be resolved easily enough, but M69 is compact (Class V on the Shapley–Sawyer scale) and cannot be resolved to the core except with the very largest telescopes.

NGC 6681 (M70)

M70 is in many ways a twin of globular cluster M69 (NGC 6637). Both are in the same constellation, Sagittarius, and both are relatively small in apparent size, M70 being about 8.0 arcminutes in diameter compared with 8.3 arcminutes for M69. Of the two, M70 is very slightly bigger, at about 68 light years in diameter compared with 42 light years for M69, and consequently M70 is a little bit brighter than M69 as well.

M70 is rated as a Class V object on the Shapley–Sawyer scale, an average sort of density. Like other globular clusters of this type, only the outer whorls of stars can be resolved easily with medium the large aperture telescopes. Interestingly, M70 is about the same distance from Earth as M69, about 29,000 light years, and in fact the globular clusters are quite near to one another, the two being separated by less than 2,000 light years.

NGC 6694 (M26)

M26 is a loose open cluster in Scutum, and one of the more overlooked Messier objects. To be fair, it isn't a particular exciting object, and it's 90 or so members do not stand out strongly against the background star field. But it can look rather nice through a low power, wide field eyepiece. It has a width of about 15 arcminutes and contains at least 90 stars.

M26 is about 5,000 light years away and believed to be about 22 light years in diameter. Unusually for an open star cluster the distribution of stars isn't greatest at its center. The likely explanation for this is that there are clouds of dust in between us and M26, and these clouds are blocking out some of the stars.

NGC 6715 (M54)

M54 is a globular cluster in Sagittarius. It lies about 87,000 light years away from Earth, making it one of the more distant globular clusters. In fact this globular cluster belongs to the Sagittarius Dwarf Elliptical Galaxy, a small satellite galaxy that orbits the Milky Way. That being the case, when Charles Messier discovered this globular cluster, he was actually discovering the first globular cluster outside the Milky Way Galaxy as well. Of course he didn't know that at the time, and it wasn't until the 1990s that the true nature of M54 was actually realized.

Given that it is so far out, its comparatively large size and brightness might be a surprise. In terms of size, it has an angular diameter of 12 arcminutes, making it quite a bit bigger than M69 (NGC 6637) and M70 (NGC 6681), two globular clusters in the same part of Sagittarius. It's also slightly brighter than those two globular clusters as well. As you'd expect given these data, M54 is intrinsically bright and large, and believed to be about 300 light years in diameter, which is twice the diameter of the Great Globular Cluster in Hercules, M13 (NGC 6205).

M54 is a very dense globular cluster and rated as a Class III object on the Shapley–Sawyer scale. It cannot be resolved to the core, and even medium to large aperture telescopes will resolve little more than a few of the outermost whorls of stars. Most observers will simply see a moderately bright, distinctly granular ball of light. Unfortunately, M54 is too far south to be properly seen from far northern latitudes; observers in southern England will find that it does not rise more than 8° above the southern horizon.

NGC 6809 (M55)

One of the less well known globular clusters in Sagittarius, M55 is a relatively southerly object and consequently difficult or impossible to see from far northern locations. Observers in southern England for example will find that it rises no more than about 7.5° above the horizon. It's a pity this globular cluster isn't an easier target because it's one of the most attractive globular clusters in the Messier catalog.

Through the telescope it is immediately apparent that this is a very loose globular cluster. On the Shapley–Sawyer scale it is regarded as a Class XI object. While it can be seen easily enough through small telescopes, M55 is a particularly attractive sight through medium to large aperture telescopes where its looseness allows a good deal of resolution to be obtained at medium to high magnifications.

M55 is between 17,000 and 18,000 light years away and has an angular diameter of about 19 arcminutes, implying an actual diameter of around 100 light years. In common with globular clusters generally, it contains mostly very old stars, and is estimated to be about 12.4 billion years old.

NGC 6818 (Little Gem Nebula)

NGC 6818 is a small but bright planetary nebula in Sagittarius. Although a relatively southerly object for observers in the northern hemisphere, it is located towards the northern edge of Sagittarius close to the border with Capricornus, and isn't too difficult to see even as far north as southern England, where it gets to more than 24° above the horizon.

NGC 6818 has an angular size of only 39 arcseconds, and can easily be mistaken for an out-of-focus star. The use of a narrowband or O-III filter should remove any doubt as to its identity, though being very small, even with filters little is revealed in terms of detail through ordinary telescopes.

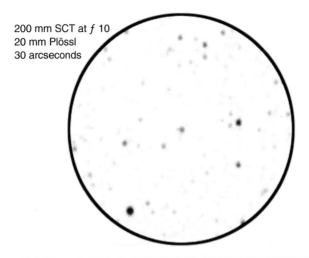

200 mm SCT at ƒ 10
20 mm Plössl
30 arcseconds

Figure 4.15. Although bright, the small angular diameter of NGC 6818 means that it can easily mistaken for an out-of-focus star. (Image produced using Starry Night Pro. AllSky data courtesy of Main-Sequence Software Inc.)

This planetary nebula is about 6,000 light years away and thought to be around 0.5 light years in diameter.

NGC 6826 (Blinking Planetary Nebula)

The planetary nebula NGC 6826 in Cygnus is a notoriously difficult object to observe with small aperture telescopes. In medium to large aperture telescopes it is an easy object, though, even yielding a little color to the view, the nebula appearing bluish-green. NGC 6826 is noteworthy for having a relatively bright star at its center. This is the dying star that created the nebula, and it shines at about magnitude 10.

Its popular name, the Blinking Planetary Nebula, comes from the fact that the central star and the nebula surrounding it seem to blink in and out of view depending on how they're being observed. Most astronomers find that when viewed directly, the star is so bright its light drowns out the nebula, making it impossible to see. But when averted vision is used, and the star isn't looked at directly, the nebula then becomes much more apparent.

You can use either narrowband or O-III filters on this object, and both will dim the central star and make the surrounding nebula easier to see. A fair bit of magnification is helpful though, since NGC 6826 has an angular diameter of only 45 arcseconds.

NGC 6826 is about 3,500 light years away and 0.6 light years in diameter. Long-Exposure photographs reveal the nebula is elliptical in shape and has two bright patches at each end of its major axis. These patches are called fast low-ionization

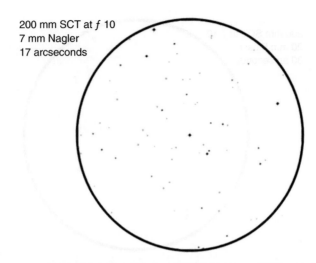

200 mm SCT at ƒ 10
7 mm Nagler
17 arcseconds

Figure 4.16. The Blinking Planetary Nebula gets its name because the nebula seems to blink in and out of view depending on whether it is being observed with averted or direct vision. (Image produced using Starry Night Pro. AllSky data courtesy of Main-Sequence Software Inc.)

emission regions, or FLIERS for short. It is thought that these are regions of hot gas that were flung out by the star sometime in the relatively recent past, perhaps a 1,000 years ago.

NGC 6913 (M29)

This distinctive little open cluster in Cygnus is an easy target for telescopes of all sizes. It contains around 50 stars, but six of the brightest form a pair of arc-shaped chains that are very close together and arranged almost like mirror images of one another.

By the standards of open clusters M29 is typical of the numerous sparse clusters that can be found scattered across the Milky Way. To be fair to M29, clouds of dust in between us and the cluster means that only a fraction of its light reaches us, with many stars that should be bright being dim or even invisible. It is thought that without this dust this cluster would be anything up to three magnitudes brighter, and were that the case, views of this cluster would be truly dazzling.

Because the intrinsic (rather than apparent) brightness of the cluster is difficult to assess, astronomers aren't absolutely sure how far away this cluster is; although it is commonly thought to be about 4,000 light years away, some astronomers believe the true value is much greater, perhaps as high as 7,200 light years. The cluster has an angular diameter of 7 arcminutes, and assuming the 4,000 light year distance, this suggests a diameter of around 10 light years.

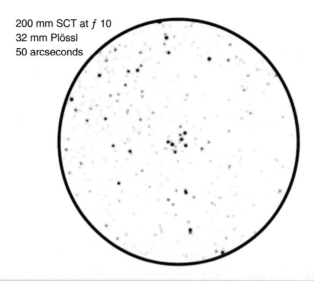

200 mm SCT at f 10
32 mm Plössl
50 arcseconds

Figure 4.17. Clouds of dust between us and M29 mean that it seems to be a sparse open cluster, even though it is actually rather bright and rich. (Image produced using Starry Night Pro. AllSky data courtesy of Main-Sequence Software Inc.)

NGC 6934

Two globular clusters in Delphinus are bright enough for amateur astronomers to see with average-sized telescopes, the other being NGC 7006. Neither are breathtaking spectacles, but they're both nice objects to try and bag while exploring the summer sky. NGC 6934 is the brighter of the two and can be seen even with relatively small telescopes, possibly even binoculars, under dark sky conditions.

NGC 6934 is rated as a class VIII object on the Shapley–Sawyer scale, which would imply it's a relatively loosely arranged cluster, but in fact resolving this object to the core is very difficult. Even with a 200-mm SCT only the outer whorls of stars will be resolved.

NGC 6934 is some 51,000 light years from Earth and thought to be about 118 light years in diameter.

NGC 6994 (M72)

M72 is the dimmer of the two bright globular clusters in the constellation of Aquarius. Far northern observers may have problems getting a good view of this object because of its location, a good 12° below the celestial equator. By contrast the other bright globular cluster in Aquarius, M2, sits virtually on top of the celestial equator, and so rises much higher above the southern horizon.

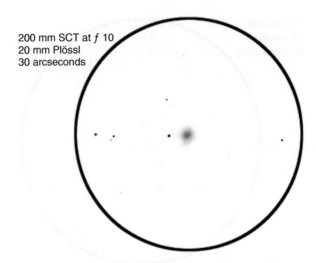

200 mm SCT at *f* 10
20 mm Plössl
30 arcseconds

Figure 4.18. NGC 6934 is a fairly bright but notoriously difficult to resolve globular cluster in the constellation of Delphinus. (Image produced using Starry Night Pro. AllSky data courtesy of Main-Sequence Software Inc.)

M72 is a nice object nonetheless, and even as far north as England it is bright enough to see even through slightly murky or light-polluted skies. It is a fairly loose cluster, and rated as a class IX globular cluster on the Shapley–Sawyer scale. Through a 200-mm SCT it appears as a somewhat faint but clearly granular blob. It is about 55,000 light years away and believed to be a little over 100 light years in diameter.

Incidentally, not far off from M72 is M73, a Y-shaped asterism of four tenth-magnitude stars that for some reason Messier included in his famous catalogue of objects potentially confused with comets. Since it isn't a true open cluster it doesn't have an NGC designation.

NGC 7006

NGC 7006 is a globular cluster in the small constellation of Delphinus, a constellation not otherwise much regarded among astronomers looking for deep sky objects. It is a very condensed globular cluster, rated at class I on the Shapley–Sawyer scale, and given its relatively small size as well, it can easily be mistaken for a planetary nebula or an out-of-focus star.

One reason NGC 7006 appears so small, about 2.8 arcminutes in diameter, is that it is a long way away, about 185,000 light years Earth. In fact NGC 7006 isn't just a long way from us; it's a long way from the center of the galaxy as well. Current estimates suggest that NGC 7006 orbits the Milky Way at a distance of about 130,000 light years from the core, making it one of our galaxy's most remote globular clusters.

Because it is relatively small and dim, this globular cluster is a challenging target for observers using small telescopes, and it may be impossible to see at all under light polluted conditions. Even with a 200-mm SCT and reasonably dark skies, NGC 7006 doesn't reveal much beyond its granularity, though some of its outer whorls of stars can be resolved.

NGC 7092 (M39)

M39 is a large, thinly populated open cluster in Cygnus that looks good in telescopes of all sizes. Because it has an angular diameter of over 31 arcminutes, it is best viewed using low powers. Wide-field instruments shouldn't have much trouble getting the cluster nicely framed against the background field of stars, but it can be a challenge seeing this cluster at its best through SCTs and Maksutov–Cassegrain telescopes. In the case of a 200-mm SCT, a 32-mm Plössl does a reasonably good job of getting the cluster in the field of view, but a much more dramatic impression will be gained by using an f/6.3 reducer-corrector along with a 2-in. wide field eyepiece such as a 35-mm Panoptic.

M39 is very nearby, only about 800 light years away, and only contains about 30 stars spread out across 7 light years of space. It is thought to be a middling age cluster less than 300 million years in age. Under dark sky conditions it is just visible to the naked eye, and although its discovery is usually attributed to either Le Gentil or Messier, it is possible that the cluster was first described by Aristotle, who mentions seeing a comet-like nebulosity in Cygnus.

NGC 7099 (M30)

This globular cluster in Capricorn is a bright and attractive example of its type, but unfortunately a bit too low in the sky for far northern observers to really enjoy (from southern England, for example, it barely gets to 15° above the southern horizon). But from locations farther south M30 is well worth tracking down and much appreciated as one of the best globular clusters for small telescopes.

M30 is a relatively dense globular cluster, rated at class V on the Shapley-Sawyer scale and consequently very difficult to resolve. It has a poor reputation among those amateur astronomers doing Messier marathons because it is usually the very last object on the list. The problem is that by the time you're ready to do M30, and M30 is somewhere above the horizon where it can be seen, it won't be long before sunrise. Consequently such marathoners are often having to find this object against a steadily brightening dawn sky!

M30 is about 26,000 light years from Earth and believed to be about 90 light years in diameter.

NGC 7209

The small constellation of Lacerta is notoriously dim and devoid of interesting deep sky objects, but it does contain a couple of reasonably worthwhile open star clusters, including NGC 7209. Visible through telescopes of all sizes, a medium to large aperture telescope will reveal a large, sparse cluster containing some 70 stars spread out across an area almost that of the full Moon.

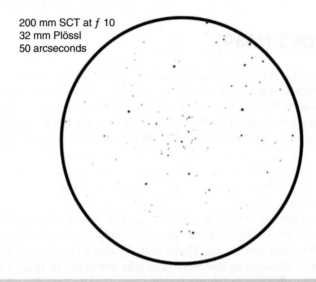

200 mm SCT at *f* 10
32 mm Plössl
50 arcseconds

Figure 4.19. NGC 7209 is a sparse, fairly dim open cluster best viewed at low powers. (Image produced using Starry Night Pro. AllSky data courtesy of Main-Sequence Software Inc.)

To get the best impression of this object a low magnification is essential. When a 200-mm SCT is being used, a 32-mm Plössl is a good starting point, but adding an f/6.3 reducer-corrector to the mix improves things significantly. Even better views will be obtained when the reducer-corrector is used with a low power, wide field 2-in. eyepiece such as a 35-mm Panoptic.

NGC 7209 is about 2,900 light years away and measures about 21 light years across. Because it contains some orange and yellow giant stars, it is thought to be quite an old cluster. Estimates suggest an age of up to 410 million years, and given how far apart its constituent stars are, the gravitational forces holding them together are probably very weak. NGC 7209 is probably an open cluster on the verge of ceasing to be one, as the stars that were born within it set off on their own trajectories across the galaxy.

NGC 7243

The second of the two notable open star clusters in Lacerta, NGC 7243 is a large, bright cluster that looks good through binoculars as well as telescopes of all sizes.

Indeed, binoculars and rich-field telescopes are probably the ideal instruments for looking at both of the Lacerta open star clusters. NGC 7209 and NGC 7243 are less than 4° apart, and when pulled into the same field of view can be seen as dim, hazy patches of light.

NGC 7243 is 21 arcminutes in diameter, and consequently low magnifications are essential. As with NGC 7209, when looking at this open star cluster with a 200-mm SCT, a 32-mm Plössl makes a good eyepiece to start your observations. If you have one, an f/6.3 reducer-corrector will improve the view by expanding the field of view, making the cluster members stand out more obviously as part of a group by adding more of the background field of stars. Ideally though, use a reducer-corrector alongside a low magnification 2-in. eyepiece such as a 35-mm Panoptic.

NGC 7243 is 2,500–3,300 light years away and contains about 40 young, hot blue stars. It is thought to be comparable in age to the Pleiades and less than 110 million years old. This makes it much younger than the nearby cluster NGC 7209.

Obscure and Challenging Deep Sky Objects

NGC 6210

NGC 6210 is a small, fairly bright planetary nebula in Hercules. While easy to see through medium to large aperture telescopes, it is very small (14 arcseconds in diameter) and consequently doesn't reveal much in terms of detail. At most, its disc-like shape will be apparent under high magnification. A narrowband or O-III filter will make the nebula easier to see, but won't make any additional detail obvious.

Photographs taken by the Hubble Space Telescope reveal that the bright part of the nebula visible through amateur telescopes is just the bright core of the structure. A much larger, but fainter, cloud of gas surrounds it. Thanks to a number of stubby extensions that protrude from the otherwise elliptical outer shell of gas, astronomers have dubbed NGC 6210 the 'Turtle in Space' (though any suggestions that this is in fact Great A'Tuin are, at this point, strictly speculative).

NGC 6210 is about 6,600 light years away and measures about 1.6 light years in diameter, though the bright core visible through telescopes is only about 0.5 light years across.

NGC 6302 (Bug Nebula)

This small planetary nebula in Scorpius is essentially invisible to observers at far northern latitudes, rising by barely a degree or so above the horizon from the point of view of observers in southern England. Observers farther south will have more success, assuming a clear horizon, because NGC 6302 is quite bright, and despite

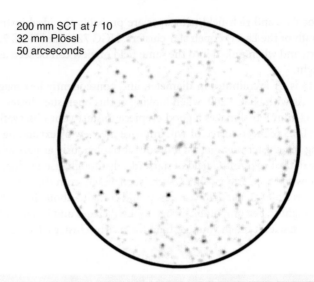

200 mm SCT at f 10
32 mm Plössl
50 arcseconds

Figure 4.20. NGC 6302 is a very small but quite bright planetary nebula, but placed too far south to be easily viewed by mid to far northern observers. (Image produced using Starry Night Pro. AllSky data courtesy of Main-Sequence Software Inc.)

only have an angular diameter of 48 arcseconds, it is reasonably easy to see. As with planetary nebulae generally, a narrowband or O-III filter will help by removing ambient light pollution and removing any difficulty distinguishing the nebula from any stars in the same field of view.

NGC 6302 has been well studied over the years. It an example of a type of nebula known as a bipolar nebula. Such nebulae have two conical clouds of gas spreading outwards from the dying star in the center, so that they look a butterfly or two eggs with their pointed ends touching. Although this is quite a common class of nebula, NGC 6302 is unusual in that we're viewing it sideways, so that the two clouds of gas are easily seen. The Ring Nebula (NGC 6720) is also a bipolar nebula, but because it's viewed end-on, we're looking down through one of the conical clouds of gas, and so we see a ring-shaped structure surrounding the central star.

NGC 6302 contains a dying star that is known to be an extremely hot object, upwards of 200,000 Kelvin. This implies that it was once a very massive star, much larger than our Sun. As is commonly the case with planetary nebulae, measurements of the distance between NGC 6302 and Earth are uncertain, though a recently proposed value of around 3,400 light years is widely accepted. That being the case, the length of the nebula would be around 2.1 light years.

Its common name, the Bug Nebula, comes from its appearance in long-exposure photographs, where it looks a bit like a bug (and a squashed bug at that). Visually, observers are more likely to see a small, somewhat elongated blur of light.

NGC 6369 (Little Ghost Nebula)

NGC 6369 is a planetary nebula in the southern part of the constellation of Ophiuchus. Visually it is small, about 30 arcseconds in diameter, and comparatively dim, disc-shaped nebula. It is difficult to see from light polluted environments, particularly from far northern locations. In southern England, this object will only be seen up to about 14° above the horizon. Narrowband and O-III filters make discerning this nebula from the background skyglow much easier.

At low powers the nebula simply looks like an out-of-focus star. Under dark sky conditions, large aperture telescopes reveal an elliptical ring shape when the nebula is viewed at high magnification. Suburban astronomers will find the ring shape much more difficult to discern, though the slightly dimmer central region may be seen to contrast with brighter outer edge of the nebula.

NGC 6369 is about 2,000 light years away. It's popular name, Little Ghost Nebula, is a reference to its relative faintness.

NGC 6531 (M21)

M21 is a young open cluster in Sagittarius that contains over 50 hot blue stars including several very hot Class B stars similar to those seen in the Pleiades.

M21 is one of the less exciting open clusters in the Messier Catalogue, but it is very close to the Trifid Nebula (NGC 6514), and the two objects taken together make a very interesting target for owners of rich field telescopes. Owners of long focal length telescopes will find getting the two objects into the same field of view a challenge. In the case of a 200 mm-SCT, an f/6.3 reducer-corrector will be essential. Used with a 32-mm Plössl, the reducer-corrector will deliver a field of view about 1° across, and that's enough to squeeze the two objects into the same field. Even better results will be had if a wide field 2-in. eyepiece can be used, such as a 35 mm-Panoptic, in which case the field of view will be twice as large, and the overall impression of the star cluster, the nebula, and the background field of stars will be even more dramatic.

There is some uncertainty over how far M21 is from Earth, but the most recent estimates suggest a distance of about 4,250 light years. M21 appears to be about 4.6 million years old, making it one of the youngest open star clusters that amateur astronomers routinely observer.

M21 is a tough object for observers in the far northern hemisphere. Observers at the latitude of Southern England will find that it barely reaches 16° above the horizon.

NGC 6613 (M18)

M18 is an open star cluster in Sagittarius. It contains up to 20 members spread out across no less than 9 arcminutes, and consequently looks sparse and comparatively poor compared with the nearby Swan Nebula (M17, NGC 6618) a mere degree or so away. Unsurprisingly, it's often overlooked by astronomers bagging the more celebrated sights in this southerly constellation.

M18 it looks best at low magnifications, when its mix of blue, orange, and yellow stars can be most easily appreciated and contrasted with the background field of stars. The cluster is fairly young, around 32 million years old, and lies about 4,900 light years from Earth and about 17 million light years in width.

Like many objects in Sagittarius it is not an easy object for far northern observers, only reaching 21° above the horizon at locations as far north as southern England.

NGC 6709

The constellation of Aquila isn't famed for its deep sky objects, despite being one of the large ones, the 20-s biggest constellation out of a total of 88, to be precise. Only two objects make it into this book, one of which is the open cluster NGC 6709 (the other is the planetary nebula NGC 6781).

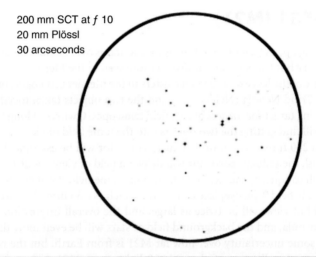

200 mm SCT at *f* 10
20 mm Plössl
30 arcseconds

Figure 4.21. NGC 6709 is a sparse, fairly nondescript open cluster. (Image produced using Starry Night Pro. AllSky data courtesy of Main-Sequence Software Inc.)

NGC 6709 contains around 60 stars, mostly blue stars, but with some yellow and orange giant stars in the mix as well. It has been generally assumed to be about 3,000 light years away and 12 light years in linear diameter, but some astronomers believe that it is a bit farther away than this, in which case its linear diameter would have to be a big greater as well.

Either way, its constituent stars suggest an age around 315 million years. This means that when this open cluster was formed, Earth was in the second half of what we call the Carboniferous Period, a time when warm, humid forests covered much of the land. Some of the wood and vegetation in these forests sunk to the bottom of swamps where it became peat, and over time, was compressed and heated to form coal. That's something to think about next time you shovel some coal onto the fireplace!

NGC 6712

This globular cluster in Scutum is not well known, but it is bright and quite easy to see. Under dark sky conditions telescopes as small as 100 mm will reveal its presence, but backyard astronomers will find telescopes in the 150–200 mm aperture range more useful. At the very least, it should be apparent as a small, granular patch of light.

NGC 6712 is notoriously difficult to resolve, despite its Class IX designation on the Shapley–Sawyer scale. At about 22,500 light years from Earth, it is closer to us than the Great Globular Cluster in Hercules, which lies about 25,000 light years away. But unlike the Great Globular Cluster, this is quite a small globular cluster. Whereas the Great Globular Cluster is 160 light years across, NGC 6712 is a mere 64 light years in diameter. So one reason it is difficult to resolve is its much smaller apparent size, 7.4 arcminutes in the case of NGC 6712 compared with 36 arcminutes for the Great Globular Cluster.

NGC 6712 is unusual among globular clusters in its proximity to the galactic core, its orbit taking it to within 1,000 light years of the absolute center of the Milky Way Galaxy. As it passes through the core, gravitational forces pull away some of its outlying stars, which go off to join the main population of stars in the Milky Way. This thins out the outer whorls of stars that surround this globular cluster, and that means that there are fewer of those stars for the amateur astronomer to resolve.

NGC 6716

This young open star cluster in the constellation of Sagittarius contains at least 75 stars, but when set against the dense star fields that make up this part of the night sky, it is very easily overlooked. It has a relatively small apparent size, about 6 arcminutes, and in small telescopes may seem to be little more than a hazy blur. Medium to large aperture telescope will reveal the individual stars that make up the cluster.

NGC 6716 is about 100 million years old and about 2,000 light years from Earth.

NGC 6779 (M56)

This globular cluster in Lyra is often overlooked, particularly given its proximity to the superb Ring Nebula (M57, NGC 6720). But while M56 isn't the best globular cluster in the sky, it is conveniently located so high up in the sky that suburban astronomers will find it a useful challenge. Most other third-rate globular clusters are much farther south in constellations such as Ophiuchus and Sagittarius, but M56 is high enough in the sky that it should be free of obstructions and less affected by light pollution that objects closer to the horizon. So if you're looking for a 'difficult' globular cluster to test out your eyes and your equipment, this is a good one to start with.

Under dark sky conditions, M56 is visible through telescopes as small as 75 mm, but you will probably need 100–150 mm to see this object under suburban conditions. Larger telescopes allow some of its stars to be resolved. M56 has an apparent size of only 8.6 arcminutes, so a fair bit of magnification will be required to see this object properly. As with all globular clusters, dark skies help, but otherwise try using averted vision if its details aren't readily apparent.

Because M56 seems to sit on a particularly rich part of the Milky Way from our point of view, it's a favorite target for observers using large aperture binoculars and rich field refractors. Owners of 200-mm SCTs will get some sense of this when low powers are used, as would be the case with a 32-mm Plössl, but for a really good view use an f/6.3 reducer-corrector and switch to a low power 2-in. eyepiece like a 35-mm Panoptic.

M56 is about 33,000 light years away and measures some 42 light years in diameter. On the Shapley–Sawyer scale it is considered a Class X object.

NGC 6781

This planetary nebula in Aquila is small, a mere 1.8 arcminutes in diameter, making it about the same size as the Ring Nebula (M57, NGC 6720) in Lyra. It is rather less well known, though, because it isn't nearly as bright, and consequently can be difficult to see under suburban conditions.

Given the right conditions and sufficient aperture (200 mm or more) this nebula can be seen to have a disc-like shape. In long-exposure photographs it actually looks more like a circular ring, but this isn't easy to see visually. A narrowband of O-III filter will help reveal its shape, though, as will averted vision.

NGC 6781 is around 5,000 light years away and has an actual diameter of about 2.5 light years.

NGC 6802

This small open cluster in Vulpecula is dim and difficult to pick out from the surrounding star field. It isn't particularly interesting in terms of how it looks, but it is small (about 3 light years across) and very old (1,700 million years), and these facts make it noteworthy. Most other small clusters are very young, and as they age, their stars drift away, eventually pursuing their own courses independently of one another, at which point the open cluster ceases to exist. Why is this open cluster different? Most likely it was once exceedingly rich, and it has in fact lose most its members, but a handful of stars, the ones we see today, are still gravitationally bound to one another.

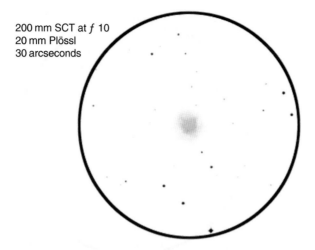

200 mm SCT at ƒ 10
20 mm Plössl
30 arcseconds

Figure 4.22. Although similar in size to the Ring Nebula, NGC 6781 is considerably fainter and more difficult to see. (Image produced using Starry Night Pro. AllSky data courtesy of Main-Sequence Software Inc.)

NGC 6802 is about 3,300 light years away from Earth. It is a mere 17 arcminutes away from the much more famous Coathanger Cluster, Collinder 399 (see 'SAO 87209' in the section on stars, below).

NGC 6819

NGC 6819 is an open cluster in Cygnus. It isn't a very highly regarded object, perhaps because it doesn't stand out too well against the rich field of stars that fills this part of the sky. But despite that, this is a nice object and well worth observing. NGC 6819 measures about 9.5 arcminutes in apparent size, and is sufficiently large and bright that it is visible through telescopes of all sizes.

NGC 6819 is about 8,200 light years away, and this means that while it is actually quite a decent size, around 18 light years in width, from our perspective it seems rather small and difficult to resolve. It is a very old open star cluster and appears to be around 2.5 billion years old.

NGC 6823

The dim open cluster NGC 6823 in Vulpecula is visible through telescopes of all sizes given good conditions, but under suburban conditions apertures upwards of 100 mm will probably be required. It is associated with a nebula, NGC 6820, but this object will not be visible from suburban situations.

200 mm SCT at ƒ 10
32 mm Plössl
50 arcseconds

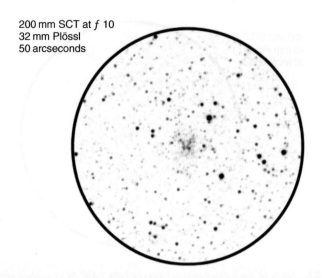

Figure 4.23. NGC 6819 is a bright open cluster, but difficult to pick out against the dense star fields in this part of Cygnus. (Image produced using Starry Night Pro. AllSky data courtesy of Main-Sequence Software Inc.)

NGC 6820 and NGC 6823 been studied by astronomers and appears to be broadly similar to the Lagoon Nebula (M8, NGC 6523) in being a site where more than one period of star formation has occurred. Taken together, the two objects are about 100 light years at their greatest width, and lie about 8,200 light years from Earth.

NGC 6864 (M75)

M75 is one of the less impressive, and consequently less often observed, globular clusters in Sagittarius. It is a small and faint object primarily because of its distance; in terms of actual size, at a diameter of about 130 light years is actually pretty respectable, and not far off the 145 light year diameter of the Great Globular Cluster in Hercules (M13, NGC 6205). But it seems so small, less than 6 arcminutes in diameter, because of the huge distance between us and M75, some 67,500 light years, more than two-and-a-half times farther away than the Great Globular Cluster in Hercules.

Besides being small and faint, M75 is also very difficult to resolve. It is a Class I object on the Shapley-Sawyer scale. So while a fair amount of magnification should show this object's granular texture, even with a large telescope not much else will be seen.

Observers at far northerly latitudes will find this object a difficult target. At the latitude of southern England the object only rises to about 16.5° above the horizon.

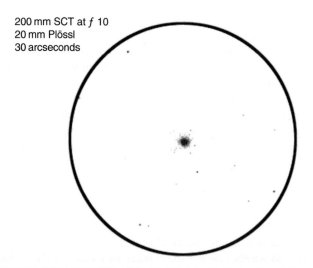

200 mm SCT at ƒ 10
20 mm Plössl
30 arcseconds

Figure 4.24. M75 is a small, highly condensed globular cluster that cannot be easily resolved, even with large aperture telescopes. (Image produced using Starry Night Pro. AllSky data courtesy of Main-Sequence Software Inc.)

NGC 6866

This open cluster in Cygnus is quite small but rich, with over 130 members spread out across about 15 arcminutes of space. It can be spotted in telescopes of all sizes, but medium to large apertures and moderate magnification are needed to resolve the majority of its members. In smaller aperture instruments only the brightest members will be seen, and being clumped together at low magnifications, it is easy to understand how when first discovered by Caroline Herschel this cluster was mistaken for a nebula.

NGC 6866 is about 3,900 light years away and measures 17 light years in diameter.

NGC 6885

NGC 6885 is a fair-to-middling quality open star cluster in the constellation of Vulpecula and centered on the bright star 20 Vulpeculae (SAO 88339). Although a rich cluster, many of its members are rather faint, and this object looks best in medium to large aperture telescopes.

It spans a patch of sky about 20 arcminutes in width, and as such, looks best in low power, wide field eyepieces. A 32-mm Plössl will do a reasonably job of bringing out the best from this cluster when used with a 200-mm SCT, but if you can use a reducer-corrector as well, and perhaps even a long focal length 2-in. eyepiece like a 35-mm Panoptic, things get even better.

In his book on the Caldwell objects, noted observer Stephen James O'Meara has written at length about the confusion the identities of objects NGC 6882 and NGC 6885, William Herschel having discovered both of them on successive nights and having written very similar descriptions for each. The problem is that the positional data given doesn't match what modern astronomers see.

In short, in the case of NGC 6882, there's no cluster to be found at the spot where Herschel described one (at the time, the NGC system wasn't invented, but to keep things simple, we'll stick with the modern designation rather than Herschel's. More confusingly, while the cluster we call NGC 6885 seems obvious enough today, Herschel didn't mention 20 Vulpeculae in its description, which is odd, given that it is the brightest star in the cluster.

Modern astronomers assume the two objects are the same thing, and star charts and planetarium programs (and indeed go-to telescope handsets) assume that this is the case. But O'Meara mentions other groupings of stars that can be made out in this region, any one of which might the clusters Herschel was describing. There's no clear-cut answer to whether or not the clusters we today call NGC 6882 and NGC 6885 are actually the ones Herschel discovered, or for that matter whether the two clusters are one and the same thing.

In any case, the object we now call NGC 6885 is about 1,950 light years away and 20 arcminutes in angular diameter, implying an actual diameter of around 11 light years.

NGC 6888 (Crescent Nebula)

NGC 6888 is a faint, difficult to observe emission nebula in Cygnus. Despite its popular name, it is actually an elliptical bubble-like structure with a major axis diameter of about 18 arcseconds. Given that it is some 5,000 light years away, this implies a maximum width of 26 light years.

The name Crescent Nebula comes from its visual appearance under dark skies. Most of the nebula is invisible, but about a quarter of the perimeter is brighter than the rest, and this can be seen under good conditions. A narrowband filter helps, and an O-III filter even more so. Large aperture telescopes are essential. If slewing the telescope to NGC 6888 doesn't help, try centering on SAO 69597; this star is just about at the center of the (comparatively!) bright arc, and if you're going to see NGC 6888 at all, this is the portion you're going to see. Try using averted vision as well.

In all fairness, this is one of the most difficult objects listed in this book, and a true benchmark for good quality skies. It isn't likely to be seen from suburban locations, but it can be seen from relatively dark exurban sites.

NGC 6910

NGC 6910 is yet another nondescript open cluster in Cygnus, one of many that can be picked out in the Milky Way by observers who are sufficiently motivated. It is actually associated with a nebula, but you have no chance of spotting that under suburban conditions.

NGC 6910 is quite bright and visible through medium aperture telescopes with ease. But it is rather small, about 7 arcminutes in diameter, so a reasonably high level of magnification will be required to fully resolve its member stars.

This star cluster is unusual in containing no fewer than four Cepheid-type variable stars. These stars have a known relationship between the frequency with which their brightness varies and their intrinsic brightness. Since we can measure how bright they seem from here, and we know how bright they actually are, astronomers can calculate precisely how far away these stars actually are. In this case, NGC 6910 is known to be about 5,400 light years away.

NGC 6940

NGC 6940 is a large, fairly bright open cluster in Vulpecula. It is about 31 arcminutes in diameter, and absolutely must be viewed with as wide a field as possible. It is really best treated as a binocular object, or else something for astronomers with rich field refractors, but if you get the required field of view with your go-to telescope, it's certainly worth adding to your observation schedule.

As with other large clusters, owners of 200-mm SCTs will find even a 32-mm Plössl limiting, and using an f/6.3 reducer-correct dramatically improves the quality of the view. Things get even better when a 2-in. eyepiece is used so that the maximum field of view is obtained. In the case of a 200-mm SCT, something like a 35-mm Panoptic will do the trick, turning in a respectable 2° field of view.

NGC 6940 contains a mix of bright blue stars and orange and yellow giant stars. It appears to be a relatively old open cluster, with an age of about a 1 billion years having been recently suggested. Given its apparent size and its distance of around 2,600 light years from Earth, NGC 6940 is estimated to be about 19 light years in diameter.

NGC 7000 (North America Nebula)

This emission nebula in Cygnus is most realistically considered a dark sky object suited to observation through rich-field refractors and short focal length reflectors. Although its visual magnitude of 5.5 suggests that it should be a bright object, it is so large that the light it emits is very spread out, so that the overall contrast of the object against the background sky is exceedingly low. Its apparent dimensions are about 2° by 1.7° meaning that it is much bigger than a full Moon (which is about 0.5° in diameter).

If you want to see NGC 7000, then there are a couple of prerequisites. The first is a light pollution filter. Although it is possible to see NGC 7000 without a filter under pitch-black skies, observers anywhere else will find a narrowband, O-III or H-beta filter dramatically improves contrast and thereby makes the nebula easier to

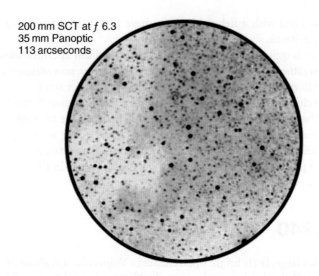

200 mm SCT at ƒ 6.3
35 mm Panoptic
113 arcseconds

Figure 4.25. NGC 7000 will likely be impossible to see from most suburban skies. One problem is its sheer size: even with a reducer-corrector and a wide-angle eyepiece, NGC 7000 overflows the field of view. The object also has an extremely low surface brightness. A narrow-band filter helps a great deal, but even so, at least fairly dark skies will be important. (Image produced using Starry Night Pro. AllSky data courtesy of Main-Sequence Software Inc.)

see. The second requirement is for a very wide field of view. In the case of a 200-mm SCT, this means the use of an f/6.3 reducer-corrector to shorten the focal length of the telescope, and a long focal length 2-in. eyepiece that then delivers the widest field of view, such as a 35-mm Panoptic. An SCT so equipped will yield a 2° field of view.

In all fairness, most suburban astronomers don't have any chance at all of seeing this object, but brave astronomers in exurban areas with excellent naked eye views of the Milky Way might feel it's worth a shot. There's certainly no way to know without trying!

NGC 7000 gets its popular name from its shape, though this is really best seen in long-exposure photographs. It is essentially a huge cloud of ionized gas more than 80 light years in diameter, though which star is producing the energy that is ionizing the gas is unknown. There is also some uncertainty as to the precise distance between NGC 7000 and Earth, though a value of 1,800 light years is widely accepted.

NGC 7027

NGC 7027 is a small but relatively bright planetary nebula in Cygnus. It is easily seen with medium to large aperture telescopes, though at low magnifications it looks like little more than a slightly blurry star. With increasing magnification it appears to be a small, blue–green nebula of vaguely elliptical (some say pillow-like) shape. A narrowband filter or an O-III filter will improve the contrast of the object, but with an angular diameter of no more than 18 arcseconds, it is very difficult to see any detail.

Although not especially well known among amateur astronomers, this nebula is much studied by professional astronomers because it appears to be extremely young. Most planetary nebulae are at the long, final stage of their life where the cooling star is surrounded by an huge, expanding bubble of ionized gas drifting off into space. NGC 7027 is different. It's at an earlier phase where the star is still very hot, having finally used up its hydrogen fuel and switched to more exotic nuclear reactions that produce a variety of heavier elements including carbon, nitrogen and oxygen. These elements collect in a layer around the dying star called a circumstellar envelope. Eventually this will get sloughed off as the gas is heated and expands by the dying star, and once ionized, it becomes the bubble of ionized gas that we see when observing most planetary nebulae.

But when we look at NGC 7027, what we see is a dying star that is only part-way through ionizing its circumstellar envelope. This phase doesn't last long, perhaps a 1,000 years or so, and the shell of ionized gas we see when looking at NGC 7027 may be as little as 500 years old. That isn't very long ago; 1509 is the year when Galileo started using a telescope!

NGC 7027 is about 3,000 light years away. It is a tiny object by the standards of deep sky objects, barely 0.25 light years in diameter.

NGC 7063

NGC 7063 is a sparse open cluster in Cygnus. It is quite easy to see, and considered a particularly worthwhile object for binoculars; it also looks attractive in telescopes of all sizes. Given its apparent size of about 7 arcminutes, a low to medium magnification works best when observing this object.

NGC 7063 is 2,300 light years away and about 4.8 light years in diameter. It is a little over 100 million years old.

NGC 7293 (Helix Nebula, Eye of God)

On paper at least the Helix Nebula sounds like it should be a fantastic target. It is one of the closest planetary nebulae to Earth, only a mere 650 light years away. Being comparatively nearby, it's also very bright, with an apparent magnitude of about 6.5, which compares favorably with that of the more famous Ring Nebula in Lyra, which only manages an apparent magnitude of 9.0. Finally, the Helix Nebula is remarkably big, its 2.5-light year diameter spanning some half a degree of space from our vantage point, about the same as the Moon.

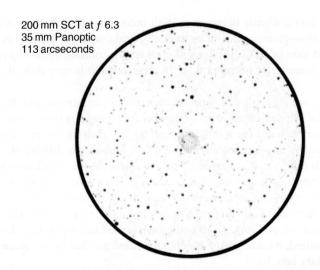

200 mm SCT at *f* 6.3
35 mm Panoptic
113 arcseconds

Figure 4.26. The Helix Nebula is a famous deep sky object, but difficult to see under suburban sky conditions. It has a low surface brightness and for mid to far northern hemisphere observers at least, does not rise very far above the southern horizon. (Image produced using Starry Night Pro. AllSky data courtesy of Main-Sequence Software Inc.)

And yet, the Helix Nebula invariably disappoints. Indeed, it often fails to show up altogether. The problem is that although it is a bright object, its big angular diameter means that the light it emits is very spread out by the time it gets to your telescope's eyepiece. So although the Ring Nebula might be dimmer on paper, it's actually a more concentrated source of light, and consequently stands out much better against the background sky. The Helix Nebula doesn't manage to do this at all, and unless your skies are pitch black, then it is likely to be completely undetectable.

Another problem for would-be observers is that the Helix Nebula has a relatively southerly location in the sky, and far northern observers will find that it never rises much above the southern horizon. This part of the sky is typically murky with dust as well as illuminated by ambient light pollution from distant cities, making it doubly difficult to catch a glimpse of this elusive nebula.

Because of its large size, a wide field of view is critical. Rich-field refractors and fast Newtonian telescopes are the instruments of choice, but a 200-mm-SCT equipped with an f/6.3 reducer-corrector and a wide field 2-in. eyepiece such as a 35-mm Panoptic works just as well.

Narrowband and O-III filters should help you see this object if it doesn't immediately spring into view, particularly if you're observing under relatively good conditions. But don't expect miracles – from his suburban observing location 25 miles north of London, the author has never found either a narrowband filter or an O-III filter much help with this object.

Colorful and Curious Stars

SAO 48796 (Delta Cygni)

Delta Cygni has a reputation for being a tough target for small to medium aperture telescopes. The problem isn't so much the fairly tight 2.7 arcsecond distance between its two components, but the difference between the brightness of the two stars.

The primary shines at about third magnitude, whereas the secondary is much fainter, at about eighth magnitude. The magnitude scale is such that magnitude 1 stars are 2.512 times brighter than magnitude 2 stars, and magnitude stars are 2.512 times brighter than magnitude 3 stars, and so on. This means that an eighth magnitude star is 100 times fainter than a third magnitude, and therein lies the rub: the secondary star is so much fainter than the primary that spotting it at all can be very tricky.

SAO 49941 (Alpha Cygni, Deneb)

Deneb is one of three bright stars that mark the corners of the what Patrick Moore calls the Summer Triangle, along with Lyra (SAO 67174) and Altair (SAO 125122). In itself it isn't a particularly attractive star beyond its brilliance, but it is a notable star.

Given its apparent magnitude of 1.3, you'd imagine it is a relatively nearby star, as is the case with Altair (at magnitude 0.8 and 16.8 light years away) and Vega (magnitude 0.6 and 20 light years away). In fact Deneb is a tremendously distant star, over 3,200 light years away, and the reason it shines so brightly is that it is intrinsically a very large and very luminous object.

Most supergiant stars are red giants, but Deneb is a white supergiant. It has a diameter at least 200 times that of our Sun. To understand just how bright Deneb is, it is useful to compare its absolute magnitude with that of the Sun. The absolute magnitude of a star is its visual magnitude when viewed from a distance of 10 parsecs (or 32.6 light years) away from it. In the case of the Sun, its absolute magnitude is a 4.8, which isn't very bright at all. From 10 parsecs away, the Sun would be only marginally brighter that the asteroid Vesta appears when viewed from Earth. By contrast, mighty Deneb would shine at an astounding magnitude –6.9, more than six times brighter than Venus appears to us at its most brilliant!

Deneb, Altair and Vega are part of a famous Chinese folktale about two lovers, a fairy princess (Vega) and a human cowherd (Altair). One of the gods was angry that an immortal and a mortal were living together, and punished them by placing them on either side of an impassable river, the Milky Way. The princess is sits on one side, and the cowherd, flanked by their two children (Beta and Gamma Aquilae) sit on the other. But once a year the magpies of the world take pity on the two lovers, and come together to form a bridge across the river, its position marked by Deneb.

SAO 67174 (Alpha Lyrae, Vega)

Vega is a very pretty, brilliantly white star. It is a an A-type main sequence star, like a lot of the brightest stars in the sky (Altair, Deneb, Fomalhaut and Sirius are all A-type stars as well). Vega is 2.7 times the size of our Sun, but it is much younger, no more than about 500 million years old. While a young star compared to the Sun (which is about 4.6 billion years old) Vega is already middle aged, and within another 500 million years will have become a red giant.

Vega is of considerable importance in the history of astronomy. In 1850 it was the very first star (other than the Sun) to be photographed, and in 1872 it was the first star (other than the Sun) to have its spectrum photographed. When Norman Pogson standardized the apparent magnitude system in 1856, he chose Vega as the zero point, giving it an apparent magnitude of precisely 0.0, though modern astronomers have tweaked the system a little, adjusting Vega's brightness to 0.03.

SAO 67315 (Epsilon Lyrae, Double–Double)

The Double–double is a very famous pair of double stars. All four members are of very similar color and brightness. At low magnifications a wide pairing will be immediately obvious, Epsilon-1 Lyrae at magnitude 6.1 and Epsilon-2 Lyrae at magnitude 5.2. These two stars separated by 3.5 arcminutes, and are easily split even with binoculars. With more magnification, each of these stars can be resolved into a tight pair of stars, Epsilon-1 Lyrae (the AB pair) separated by 2.5 arcseconds and Epsilon-2 Lyrae (the CD pair) separated by 2.4 arcseconds. Resolving the Double–double down to all four stars can be a bit of a challenge for small aperture telescopes, but is far from impossible given good optics, cooperative seeing conditions, and suitably high magnifications.

SAO 69636 (RS Cygni)

Carbon stars tend to be deep red in color, and RS Cygni is no exception. This carbon star varies in brightness dramatically though irregularly, from magnitude 6.5 down to magnitude 9.5, an almost 16-fold difference in brightness. Like other carbon stars, this variability is presumably connected to the appearance and disappearance of clouds of carbon soot on the surface.

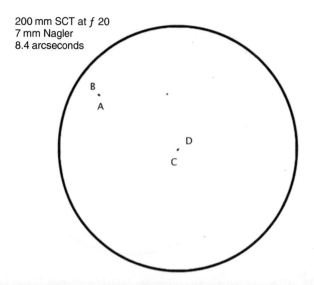

200 mm SCT at ƒ 20
7 mm Nagler
8.4 arcseconds

Figure 4.27. The Double–double can be a challenging target for small aperture telescopes but well worth the effort. (Image produced using Starry Night Pro. AllSky data courtesy of Main-Sequence Software Inc.)

SAO 72509 (8 Lacertae)

There are five stars visible in this seven-star system through medium to large aperture telescopes. Four of them form a neat little arc-shaped asterism. Components A and B are the first two in the arc and readily spotted, a pair of white stars at magnitudes 5.7 and 6.5 separated by over 22 arcseconds. The next star along the arc is component C, a magnitude 10.5 star. The final star in the arc is D, a magnitude 9.3 star. 8 Lacertae E is more than 3 arcminutes away from component A, and at magnitude 7.8, considerably brighter than the C and D components.

SAO 85648 (95 Herculis)

A celebrated double star consisting of two stars of similar brightness, a magnitude 4.9 white star and a magnitude 5.2 yellow star. Their colors are such that viewed together they resemble blobs of silver and gold. The two components are separated by about 7 arcseconds, and they are easily split even with medium aperture telescopes.

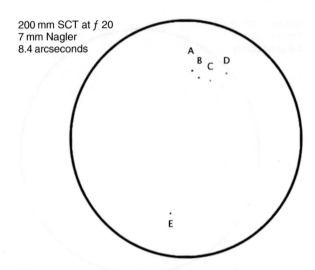

Figure 4.28. 8 Lacertae is a remarkably complex multiple-star system, but not too difficult to resolve with telescopes of medium to large aperture. (Image produced using Starry Night Pro. AllSky data courtesy of Main-Sequence Software Inc.)

SAO 85753 (100 Herculis)

This is another beautiful double star in Hercules, this time pair of magnitude 5.8 blue–white stars. The two components are 14 arcseconds apart, making them a particularly good target for small aperture telescopes.

SAO 87036 (2 Vulpeculae)

2 Vulpeculae a double star system with a (very slightly variable) magnitude 5.5 primary and magnitude 9.5 secondary. The two components are separated by only 1.8 arcseconds, and the steep difference in brightness makes 2 Vulpeculae a particularly difficult split for small to medium aperture telescopes.

SAO 87209 and the Coathanger Cluster

This star is at the center of a famous asterism in Vulpecula known as the Coathanger Cluster or Collinder 399. As its popular name suggests, this asterism looks like a coat hanger, consisting as it does of a long chain of stars with a hook-shaped chain of stars attached to the middle.

For a long time this cluster was assumed to be an open cluster, but it is now known that this is not the case, and the Coathanger Cluster is simply a chance alignment of stars. The ten bright stars that make up the coat hanger part of the cluster are traveling in different directions and lie at varying distance from Earth. Asterism though it may be, this cluster remains a great favorite among amateur astronomers, particularly those participating in star parties or showing a few celestial sights to family members.

Like the Pleiades, this is one of those objects that looks best in binoculars and rich-field refractors. Depending on which stars your mind's eye includes in the cluster, it covers something like 1–1.5° of space, so a very wide field is necessary to see the whole thing at once. In the case of a 200-mm SCT, an f/6.3 reducer-corrector and a very wide field 2-in. eyepiece like a 35-mm Panoptic will be required.

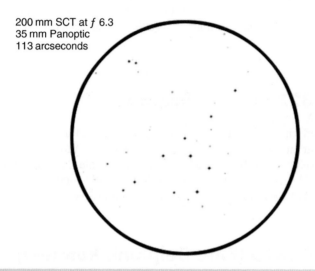

200 mm SCT at f 6.3
35 mm Panoptic
113 arcseconds

Figure 4.29. Arguments persist among professional astronomers as to whether the Coathanger Cluster is a true open cluster or merely a chance alignment of stars, but it is a popular target for backyard astronomers nonetheless. (Image produced using Starry Night Pro. AllSky data courtesy of Main-Sequence Software Inc.)

SAO 87301 and 87302 (Beta Cygni, Albireo)

Probably the best-loved double star in the sky thanks to its brightness, the ease with which is located, and the vividness of its colors. Although widely treated as a summertime object, like the other objects in Cygnus this double star is also well placed right through autumn and into early winter. Consequently Albireo is a double star that many amateur astronomers find themselves returning to again and again.

The Albireo system contains a third-magnitude yellow giant star (the primary) and a fifth-magnitude blue dwarf star (the secondary). The primary star is itself a double star, but its two components are so close together they cannot be resolved through telescopes.

Although Albireo looks good through telescopes of all sizes, it is a particularly nice object when viewed through small to medium aperture instruments. In large telescopes the sheer brightness of the two stars tends to wash out the color contrast a bit.

SAO 88098 (16 Vulpeculae)

This difficult double star consists of a magnitude 5.8 primary and magnitude 6.2 secondary only 0.9 arcseconds apart. Although the two stars are of similar brightness, their closeness means that splitting this double is a real challenge even with 200 mm aperture telescopes.

SAO 88276 (Theta Sagittae)

This is a good triple-star system for small to medium aperture telescopes, though the C component (at magnitude 7.3) is quite a distance from A to B (magnitudes 6.3 and 8.6 respectively). A and B are 11 arcseconds apart, while C is about 1.5 arcminutes away, so you'll need to play around a bit to get the right combination of magnification and field of view to see all three stars at the same time.

SAO 106316 (Beta Delphini, Rotanev)

A notoriously challenging double star for large aperture telescopes. It consists of a pair of magnitude 4.0 and 4.9 yellow stars separated by a mere 0.6 arcseconds. Apertures upwards of 200 mm are needed to split this tough double star.

Alpha (SAO 106357) and Beta Delphini possess most peculiar names, Sualocin and Rotanev. These names appeared with explanation in the 1814 edition of the *Palermo Catalogue* of bright stars. This catalog was compiled mostly by a junior astronomer by the name of Niccolò Cacciatore, at the time working under Giuseppe Piazzi (who is perhaps most famous today for his discovery of the asteroid Ceres).

Over the years various theories were put forward to explain the names, most notably by Frances Rolleston in her book *Mazzaroth or The Constellations*. She was interested in the names of the stars and their biblical and esoteric significances, and believed that the names were Scalooin and Rotaneb, the former from the Arabic for 'swift' and the latter the Syrian or Chaldean for 'swiftly running water.' The English astronomer Thomas William Webb solved the mystery, realizing that when the two names are written backwards they spell Nicolaus Venator, the Latinized version of Niccolò Cacciatore.

SAO 159764 (Nu Scorpii)

Nu Scorpii is often described as the southern-sky counterpart to the Double–double in Lyra, but even in Southern England it gets a respectable 16° above the southern horizon. It is a rather more difficult double star to completely resolve than the Double–double though. Even at fairly low powers, a small aperture telescope should be able to show two fairly bright stars about 40 arcseconds apart. With more magnification these two stars can be resolved into two pairs. A small aperture telescope should split the fainter pair, a pair of magnitude 6.8 and 7.5 stars 2.3 arcseconds apart. However, a medium aperture telescope is needed for the brighter pair, magnitude 4.1 and 6.2 stars just 1.3 arcseconds apart.

SAO 163614 (Rho Capricorni)

Rho Capricorni consists of a difficult double within an easy double. The easy part is separating the yellow magnitude 5.0 primary (called Rho Capricorni A) from the orange magnitude 6.6 secondary (Rho Capricorni D). This can be done with binoculars, the two stars being separated by well over 4 arcminutes. However, the primary also has a magnitude 10 companion 1.3 arcseconds away, and splitting these two stars is notoriously difficult. Besides a large aperture telescope, you'll also need steady skies and good eyesight. Because of its relatively southerly declination, this is an especially tough challenge for mid to far northern observers.

SAO 163626 and 163625 (Omicron Capricorni)

Omicron Capricorni is a classic double star consisting of a nicely matched pair of magnitude 5.9 and 6.7 stars. They are both white in color and separated by about 22 arcseconds. They are easily split at medium to high magnifications even through relatively small telescopes.

SAO 184415 (Alpha Scorpii, Antares)

The ancient Greeks called their war god Ares, and that was the name they gave to the planet we know by its Roman name, Mars. Like the other planets, Mars moves along a band of sky known as the ecliptic, and sometimes it passes by a star in the constellation of Scorpius that was almost as bright as Mars and just as red. This star became known as Antares, the 'rival of Ares.'

Antares is a challenging double star. At magnitude 0.96, the primary is one of the brightest stars in the sky, though its southerly declination means that far northern observers rarely get to see this star in its fully glory, and instead have to make do with seeing it twinkle through the haze close to the horizon. But if you can get a good view of Antares, it's worth making an effort to spot its magnitude 5.4 companion star about 2.6 arcseconds away. In theory this distance is sufficiently generous that this double star should be an easy split. But in practice the two stars are so different in brightness that splitting them is very difficult. The minimum aperture required is around 150 mm, and this assumes steady skies and high levels of magnification.

SAO 184382 (Rho Ophiuchi)

Rho Ophiuchi is an interesting double star system associated with a reflection nebula. The magnitude 5.1 and 5.7 yellow stars are separated by only 3.1 arcseconds, but because they are similar in brightness they are easy to split, even with a small aperture telescope. The nebula is, however, difficult to see, particularly from suburban locations.

SAO 211117 (RY Sagittarii)

RY Sagittarii is a variable that varies between a maximum magnitude of 5.8 and a minimum magnitude of 14. It is a variable star as the R Coronae Borealis type, a supergiant that periodically (and unpredictably) drops in brightness across a few weeks, and then takes months to brighten back up again. It is believed that carbon periodically condenses out to form patches of soot on the surface of the star that blocks visible light, dimming the star. When the soot is dispersed, the star regains its normal brightness. But why the soot is formed, how it ends up on the surface of the star, and what gets rid of it all remain mysteries.

SAO 225426 (Pi Lupi)

Lupus is a southerly constellation and accessible only to observers at low to mid northern latitudes. It does contain several interesting double stars though, including Pi Lupi, a relatively close (1.6 arcseconds apart) pair of almost identical white stars (one at magnitude 4.7, and other at magnitude 4.8).

SAO 225638 (Mu Lupi)

Mu Lupi is a good triple star system for observers at low to mid northern latitudes. The AB pair is a very tough split, the magnitude 5.1 and 5.2 components being barely 1 arcsecond apart, and a stiff challenge for telescopes of medium aperture. The C component is much easier to resolve; a magnitude 7.2 star almost 28 arcseconds away from the AB pair.

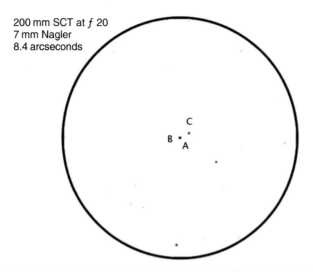

200 mm SCT at *f* 20
7 mm Nagler
8.4 arcseconds

Figure 4.30. Mu Lupi is a tough triple-star system, with components A and B being very difficult to separate. (Image produced using Starry Night Pro. AllSky data courtesy of Main-Sequence Software Inc.)

SAO 225938 (Gamma Lupi)

Gamma Lupi is an extremely tough double star that can only be split with large aperture telescopes. The stars orbit each other every 190 years, and at their widest the angular separation is only 0.8 arcseconds. They are currently slightly closer together than this, and will only reach their greatest separation in 2014. They are both blue–white stars and very similar in brightness, one at magnitude 3.5, the other at magnitude 3.6. Because of the southerly declination of Gamma Lupi, this double star is an especially difficult target for mid latitude observers, and will be impossible for anyone as far north as southern England simply because it never rises above the horizon.

200 mm SCT at / 20
7 mm Nagler
8.4 arcseconds

Figure 4.30: Maximum field of view you can see with an eyepiece. Around 8 times an object to resolve. Image produced using Sky Tools Pro Astro Ability, proprietary software program.

SAO 225938 (Gamma Lupi)

Gamma Lupi is an extremely tough double star that can only be split with large aperture telescopes. The stars orbit each other every 190 years, and at their widest the angular separation is only 0.8 arcseconds. They are currently slightly closer together than this, and will only reach their greatest separation in 2016. They are both blue-white stars and very similar in brightness, one at magnitude 3.5, the other at magnitude 3.6. Because of the southerly declination of Gamma Lupi, this double star is an extremely difficult target for mid-latitude observers, and will be impossible for anyone as far north as southern England simply because it never rises above the horizon.

CHAPTER FIVE

Autumn

For observers in the northern hemisphere, autumn is a difficult season for deep sky observing. The Moon follows a path at this time of year that means while waxing and waning gibbous it is far above the horizon flooding the night sky with its light. Although appreciated in the past as a source of light for farmers and hunters working in their fields and forests, those nights when the Harvest Moon and Hunter's Moon are shining brightly are not good times to go looking for deep sky objects.

Compared with summer, autumn evenings are colder, and astronomers will need to take this into consideration before they go outside. To start with they'll need to put their telescopes outside for half an hour or more before they are planning to start observing. If that isn't done, warm air inside the tube will interfere with the light passing through it, reducing image quality. Because it's colder, astronomers need to take care to wrap up warmly before they go outside.

These inconveniences aside, autumn does offer up some great deep sky observing opportunities. Many of the best summertime objects remain on view well into the season, the notable constellations of Cygnus and Lyra in particular staying high in the sky deep into November. For most of the season, the constellation of Pegasus is easy to spot, and with it comes the globular cluster M15, one of the best globular clusters available to northern hemisphere observers.

Autumn is a good time to observe the Milky Way as it passes through the circumpolar constellations, in particular Cassiopeia and Perseus. When we look at the Milky Way in this part of the sky we are looking at the Perseus Arm of the Milky Way Galaxy. This is the arm of the galaxy outside our own Local Spur, so instead of looking towards the galactic core (as we did in summer) we are looking away from the core. So while the Milky Way is broad and obvious, it looks dimmer at this time

N. Monks, *Go-To Telescopes Under Suburban Skies*, Patrick Moore's Practical Astronomy Series, DOI 10.1007/978-1-4419-6851-7_5,

of year because it contains fewer stars: we're looking at a spiral arm that's thinning out towards the edge of the galaxy, as opposed to a spiral arm with the galactic core as a backdrop!

The way we're viewing the Milky Way at this time of year is also why globular clusters aren't numerous. Most of the globular clusters associated with the Milky Way Galaxy swarm around its core, but because we aren't looking at the core, we don't see many of those globular clusters. On the other hand, there are still lots of open star clusters, including some of the very best of them. The Perseus Double Cluster (NGC 869 and NGC 884) is perhaps the most celebrated, and a superb object by any standards, but there are lots of others. There are some nebulae and galaxies to be seen this season as well, but on the whole these are moderately difficult to challenging ones.

Showpiece Objects

NGC 457 (ET Cluster, Kachina Doll Cluster, Owl Cluster)

NGC 457 is one of the best known open clusters not included in the Messier Catalogue. It is a great sight through telescopes of all sizes, though small to medium aperture telescopes might be the most fun. When viewed through such telescopes, two bright stars will immediately spotted; there are invariably seen as the eyes of a figure. Below them, a dense cluster of stars may be interpreted as the figure's body, and then two rows of somewhat bright stars on either side of the cluster become the figure's arms or wings.

Many astronomers have likened NGC 457 to an owl with its bright eyes staring at the observer, and its wings lifted, as if ready to fly from its perch. Others see a Kachina Doll, a religious artifact made by Hopi Indians. But astronomers of a certain age will probably see ET, the Extraterrestrial, with his bright eyes, compact body, and long arms.

NGC 457 has an angular diameter of about 13 arcminutes. It is about 9,000 light years away and 34 light years in diameter. The brighter of the two 'eyes' is Phi Cassiopeiae (SAO 22191), an F-type supergiant some seventy times the diameter of the Sun. Despite appearances to the contrary, this star is not a member of the cluster NGC 457 at all, and only happens to be in the same line of sight. Phi Cassiopeiae is in fact much closer to Earth, at a mere 2,330 light years away.

NGC 869 and NGC 884 (Perseus Double Cluster)

The two open clusters that make up this object are so close together that they're invariably observed at the same time. They are visible to the naked eye as a faint,

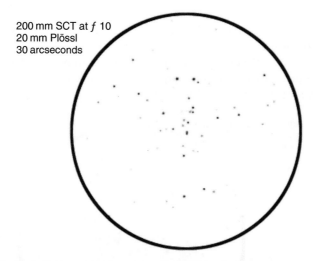

200 mm SCT at ƒ 10
20 mm Plössl
30 arcseconds

Figure 5.1. NGC 869 is a lovely open cluster with a distinctive shape described in lots of different ways. (Image produced using Starry Night Pro. AllSky data courtesy of Main-Sequence Software Inc.)

elongated patch of light halfway between Cassiopeia and Perseus, and were known at least as far back as the time of the Greek astronomer Hipparchus. Ptolemy also mentions them, though both astronomers referred to them as single objects: only with the invention of the telescope would their dual nature be revealed.

NGC 869 is sometimes known as h Persei and NGC 884 as chi Persei. These are names of a type normally used for stars. The noted variable star Algol, for example, is beta Persei. So why were the two halves of the Double Cluster given star names? Because until the eighteenth century, it wasn't clear to astronomers what these objects were. Like many deep sky objects, they were simply described as 'clouds' or 'cloudy stars,' i.e., nebulae in Latin. In 1788 William Herschel correctly described them as clusters of stars, and from then onwards, these two open star clusters have remained firm favorites among northern hemisphere astronomers.

NGC 869 is slightly closer to Earth than NGC 884, the two objects being 7,100 light years and 7,400 light years away, respectively. Both are about 30 arcminutes in diameter, implying actual diameters of about 62 light years for NGC 869 and 65 light years for NGC 884. They are both very rich clusters, containing well over 300 stars each.

Both are very young clusters. NGC 869 is 5.6 million years old and NGC 884 is 3.2 million years old. Together with the fact that they are physically near to each other probably means that they were both born from the same star-forming nebula.

The Perseus Double Cluster is a difficult object to appreciate in long focal length telescopes including most SCTs and Maksutov–Cassegrain telescopes. It looks best in instruments that can offer a field of view of at least 4°. Even with an f/6.3 reducer-corrector and a wide field 2-in. eyepiece like a 35-mm Panoptic, a 200-mm SCT is limited to a 2° field of view. That's enough to get the cores of both clusters into the field of view, but the visual impact isn't nearly the same as that from a fast reflector or rich field refractor.

Still, even if you have to view them individually, both NGC 869 and NGC 884 are first rate open clusters. Just don't forget to have a peek at them through a pair of binoculars while you're outside.

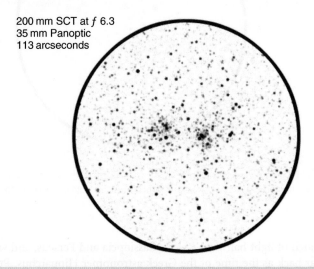

200 mm SCT at ƒ 6.3
35 mm Panoptic
113 arcseconds

Figure 5.2. The Perseus Double Cluster is a spectacular object that needs to be viewed through wide-field eyepieces.(Image produced using Starry Night Pro. AllSky data courtesy of Main-Sequence Software Inc.)

NGC 1432 (M45, Pleiades, Seven Sisters)

Almost certainly the most famous open cluster in the northern sky, the Pleiades need virtually no introduction. Suffice it to say that this open star cluster in the constellation of Taurus is one of the biggest and certainly the brightest of its type in the sky. The Pleiades is easy to see from all but the most light-polluted locations, and they look superb through even binoculars.

The Pleiades are about 440 light years from Earth. The brightest members of the cluster – the so-called Seven Sisters – span about a degree of space, but there are numerous other members lying further outside this core area. All told, more than 1,000 stars have been confirmed as members of the cluster.

As an aside, most people looking at the Pleiades will see only five or six stars with the naked eye. The name Seven Sisters more likely comes from a myth or story, such as that of the seven daughters of the Titan Atlas, that was subsequently applied to cluster. In other words, the cluster wasn't named after any seven mythological objects or people because there were seven clearly visible stars.

The age of the Pleiades is a matter of some debate, though there is a general consensus that they are a comparatively young cluster, with some estimates of their age being as low as 20 million years. In recent years this has been revised upwards considerably, to 100 million years or more.

Like the Perseus Double Cluster (NGC 869 and NGC 884) the Pleiades looks best through low power, wide-field instruments. Binoculars are probably the best instruments of all, or failing that, fast reflectors and rich-field refractors do a good job too.

The problem is that the cluster spans a whopping 2° of space, and that simply isn't going to fit into the field of view of most SCTs and Maksutov–Cassegrain telescopes equipped with 1.5-in. eyepieces. The maximum field of view for a standard 200-mm SCT using 1.5-in. eyepieces is about 50 arcseconds, and even with an f/6.3 reducer-corrector added to the system, that only goes up to about 75 arcseconds. Two-in. eyepieces help somewhat, with the maximum field of view being raised to about 2°, but while that means the Pleiades can be squeezed into the field of view, the result may be disappointing. There are plenty of bright stars to see, but without the empty background field, the specialness of the cluster is lost.

Under very dark skies, a nebula can be made out around some of the stars, most readily Merope (SAO 76172). This reflection nebula is easily overwhelmed by any light pollution, and essentially impossible to see from suburban sites. In the past it was assumed that this nebula was dust leftover from the formation of the star cluster, but it is no longer certain that this is the case. Instead it appears that the Pleiades is simply passing through a dusty part of the Milky Way, and in doing so, some of its starlight is reflected and appears to us on Earth as a nebula.

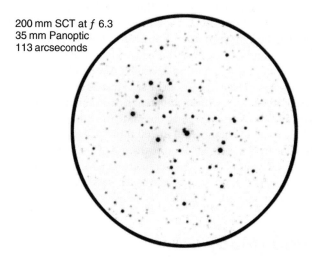

200 mm SCT at *f* 6.3
35 mm Panoptic
113 arcseconds

Figure 5.3. Although a spectacular object when viewed through a rich field refractor or binoculars, the Pleiades isn't an object that looks particularly good through an SCT. (Image produced using Starry Night Pro. AllSky data courtesy of Main-Sequence Software Inc.)

NGC 1514

NGC 1514 is a small but bright planetary nebula in the constellation of Taurus. It is small, with an angular diameter of about 2 arcminutes, and looks very much like an out-of-focus star. But while it isn't particularly special in terms of its appearance, but it is exceptional in having a very bright star at its center.

Planetary nebulae are of course formed by dying stars, and this generally implies that the central star is rather dim. In the case of the Ring Nebula (M57, NGC 6720) for example the central star has a magnitude of just 15.3. But NGC 1514 is different. Its central star (SAO 57020) glows at a very respectable magnitude 9.4, and as such, is well within the grasp of any medium to large aperture telescope. In fact the nebula part of this object is more difficult to see than the central star!

NGC 1514 is about 600 light years away, and the nebula is about 0.4 light years in diameter.

NGC 1912 (M38)

The constellation Auriga contains three showpiece open clusters, M38, M37, and M36. Of the three of them, M38 is the more distant, and although it contains something like 120 stars, it is still difficult to resolve through small aperture telescopes. Using such instruments it tends to look like a fuzzy patch of light with just a few resolved stars. As aperture increases, more stars become resolved, but the center of the cluster remains fuzzy.

M38 is about 21 arcminutes in diameter, corresponding to an actual linear diameter of about 25 light years. Given its size, this is an object best examined under low magnifications. For a 200-mm SCT, a good eyepiece would be a 32-mm Plössl.

M38 is about 4,200 light years away. It contains a few red giants, visible as yellow–orange stars that stand out against the usual blue–white ones, and is therefore believed to be of middling age, around 220 million years old.

Another open cluster, NGC 1907, can be seen just 30 arcminutes away from M38. It is much smaller than M38, with an apparent size of 7 arcminutes. The two clusters look especially attractive if observed at the same time, but that will require a field of view of 2° or more. If a standard f/10 200 mm-SCT is being used, getting them both into the same field if view will require the use of an f/6.3 reducer-corrector and a suitably long focal length 2-in. eyepiece, such as a 35-mm Panoptic.

NGC 1907 contains at least 30 stars scattered across 9 light years. It is about 4,500 light years away from Earth.

NGC 1960 (M36)

M36 is the loosest of the three Messier catalog open clusters in Auriga. Measuring about 12 arcminutes in diameter and containing around 60 stars, it is an easy object for binoculars and small telescopes. Medium the large aperture telescopes will reveal more stars, and the cluster is easily resolved to the core. It contains many hot, young B-type stars, and has an estimated age of just 25 million years.

M36 is about 6,300 light years away. Assuming this distance, its 12 arcminute angular diameter implies an actual diameter of about 14 light years.

NGC 2099 (M37, Salt-and-Pepper Cluster)

Of the three showpiece clusters in Auriga, M37 is notably more difficult to resolve than either M36 or M38. So while often said to be the most beautiful of the open star clusters in the far northern sky, it's a difficult object for owners of small aperture telescopes. Indeed, this open cluster is so nebulous when viewed through small aperture telescopes that it might even be mistaken for a loose globular cluster!

Medium to large apertures should remove any doubt as to its type, but this is still a cluster to spend a little time examining. A low magnification level is preferable here since the cluster spans 24 arcminutes, but with care it should be apparent that this open cluster contains quite a few orange stars among the expected blue–white ones. These orange stars are in fact red giants, and indicate that this open star cluster is quite mature, and old enough for some its member stars to be moving into the final stages of their lives. Current estimates place the age of M37 at about 300 million years.

M37 is about 6,200 light years away and about 30 light years in diameter. It contains at least 150 stars, including more than 12 red giants. This is one of several clusters to which the name Salt-and-Pepper Cluster has been applied. At least two

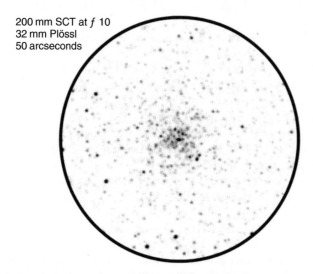

200 mm SCT at *f* 10
32 mm Plössl
50 arcseconds

Figure 5.4. M37 is a tricky open cluster, bright but quite difficult to resolve to the core. (Image produced using Starry Night Pro. AllSky data courtesy of Main-Sequence Software Inc.)

other Messier catalog clusters have had this name given to them as well, M11 (NGC 6705) and M52 (NGC 7654).

NGC 7009 (Saturn Nebula)

This bright planetary nebula in Aquarius is oddly overlooked by many astronomers, perhaps because it isn't in the famous Messier catalog. It is also notoriously difficult to find using star-hopping methods, but that isn't something owners of go-to telescopes need to worry about! Indeed, the Saturn Nebula is rather easy object to observe once located, even in somewhat light polluted skies. Like all small planetary nebulae the Saturn Nebula can be mistaken as an out-of-focus star, but the use of a narrowband or O-III filter should eliminate any such confusion.

Small telescopes reveal the Saturn Nebula has an elliptical rather than circular shape, another clue that this is a nebula rather than a star (or planet, for that matter). Larger apertures, from 200 mm upwards, reveal a pale green coloration and two odd extensions to the tips of the edges of the nebula, known as fast low-ionization emission regions, or FLIERS for short. These are often described as looking a bit like the rings around Saturn, and hence the common name for this object. They are in fact the brightest portions of two jets of hot gas emanating from the nebula.

The distance between Earth and the Saturn Nebula is not precisely known, but is believed to be about 3,000 light years. Given its angular size is about 0.5 arcminutes, this implies the nebula is in fact a little over 0.4 light years in diameter.

NGC 7078 (M15, Great Pegasus Cluster)

M15 is a fine globular cluster and a real treat for observers from late summer through to early winter. It is just about visible to the naked eye under dark skies as a hazy patch, and easily visible through binoculars. Telescopes of all sizes deliver lovely views, but as always with globular clusters, a reasonably large aperture is necessary if you want to start resolving some its stars.

One reason M15 looks so nice is that it is quite condensed, rated as a class IV object on the Shapley–Swayer scale that runs from the most condensed at I to the least condensed at XII. Very condensed globular clusters are virtually impossible to resolve through the sorts of telescopes owned by amateur astronomers, but M15 is condensed enough it looks good, but has just enough looseness to it that medium-aperture telescopes (from 150 mm upwards) will begin to resolve the outer whorls of stars.

Try using a fairly high magnification when viewing this object. Some observers reckon magnification levels 1.5 times the aperture of the telescope (in mm) is what's required. On these brighter globular clusters, the author finds magnification levels of 0.75–1 times the aperture of his 200-mm SCT work nicely, but everyone's eyes are different, so it pays to experiment a bit with whatever eyepieces and/or Barlow lenses you happen to have.

M15 is about 33,600 light years from Earth and is believed to be about 175 light years in diameter. M15 is unusual in being one of the densest globular clusters known in the Milky Way Galaxy. In fact it has an extraordinarily dense core, with about half its mass concentrated in a spherical region at the center only 20 light years in diameter. It isn't clear whether this is because the stars inside the core are tightly packed, or else because M15 contains an extremely massive object at its core, such as a supermassive black hole.

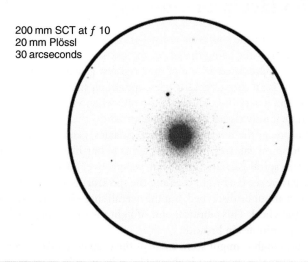

200 mm SCT at *f* 10
20 mm Plössl
30 arcseconds

Figure 5.5. M15 is a superb globular cluster, and a showpiece object during the first half of the autumn. (Image produced using Starry Night Pro. AllSky data courtesy of Main-Sequence Software Inc.)

NGC 7654 (M52)

M52 is one of the best of the many open clusters in Cassiopeia. It is a relatively young cluster, around 35 million years old, containing more than 200 hot blue–white stars. At low magnifications the cluster appears very busy, and the core may be difficult to resolve. A low to medium magnification will help to resolve more of the stars in the center of the cluster. For a 200-mm SCT, try using a 20-mm Plössl or better yet, a medium magnification, wide-field eyepiece such as a 13-mm Nagler.

M52 is about 5,000 light years distant and about 19 light years in diameter. The bright (magnitude –8.1) yellow star near the center of the object is not a member of the cluster, but merely in the same line of sight.

A little over 0.5° from M52 is a nebula, NGC 7635, centered on the on the magnitude –8.4 star SAO 20575. Part of this nebula forms a bubble-like shape known as the Bubble Nebula, a notoriously difficult to see object. But other parts of the nebula are brighter and more readily seen (under dark skies) than the Bubble itself, particularly with large aperture telescopes equipped with narrowband filters. Although unlikely to be accessible to observers under suburban skies, exurban astronomers may find NGC 7635 worth a shot. O-III and H-beta filters are helpful.

Interesting Deep Sky Objects

NGC 253 (Sculptor Galaxy)

This famous galaxy in the southerly constellation of Sculptor is noted among professional astronomers for being the site of intense star formation. Among amateur astronomers it is appreciated as one of the brightest galaxies, and from mid to low latitude locations is an easy target for telescopes of all sizes. Far northern observers will find NGC 253 a very difficult target though; when viewed from southern England for example it only rises 13° above the horizon.

NGC 253 is one of the intermediate spiral galaxies, galaxies that don't have clear bars like barred spiral galaxies, but do have hints of bar-like structures that set them apart from plain spiral galaxies. From our perspective, NGC 253 is viewed almost edge-on, so what we see is a bright core and the spiral arms at a severe angle. The spiral arms can just about be discerned, but the overall shape is more like a long, narrow ellipse than a pinwheel. This elliptical blur of light measures roughly 27 arcminutes in length by 7 arcminutes in breadth.

Of course although it might seem small at the eyepiece, NGC 253 is huge in real terms. Recent estimates suggest it is about 11.4 million light years away, so its angular width of 27.5 arcminutes implies a diameter of over 90,000 light years. Still, at just 11.4 million light years away this is quite a nearby galaxy. In fact NGC 253 is a member of the Sculptor Group, one of the groups of galaxies nearest to our own group of galaxies, the Local Group.

NGC 288

NGC 288 is a moderately large globular cluster in Sculptor less than 2° away from the Sculptor Galaxy, NGC 253. Needless to say, if you're having trouble seeing the Sculptor Galaxy, NGC 288 won't be any easier.

This is a remarkably good globular cluster even though it is terribly well known. It is comparable to a middle rank Messier globular cluster, and easy to see through medium and large aperture telescopes. It is quite a loose globular cluster, and rated as a Class X object on the Shapley–Sawyer scale. Telescopes with apertures of 200 mm will show a small bright core surrounded by a thick whorl of outlying stars.

NGC 288 is about 13 arcminutes in diameter and about 28,700 light years away. This implies an actual diameter of about 110 light years. NGC 288 is less than 37 arcminutes away from the Southern Galactic Pole. In fact NGC 288 follows a retrograde orbit that takes it past the Southern Galactic Pole once every 200 million years.

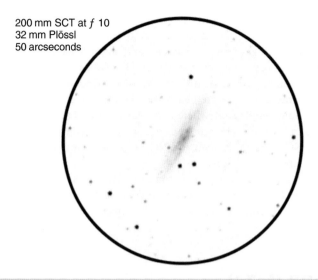

200 mm SCT at ƒ 10
32 mm Plössl
50 arcseconds

Figure 5.6. NGC 253 is a relatively bright galaxy, but its southerly location can make it a difficult target for northern sky observers. (Image produced using Starry Night Pro. AllSky data courtesy of Main-Sequence Software Inc.)

NGC 559

The constellation of Cassiopeia is well known for its open clusters, and much fun can be had perusing these at low magnifications using binoculars, rich field refractors and fast Newtonian telescopes. Most look best at low powers and through wide field eyepieces, but NGC 559 is a relatively small cluster that is well suited to observation through a go-to telescope.

In terms of size it has an angular diameter of less than 7 arcminutes, but it contains a nice mix of bright as well as faint stars, and stands up well to magnification.

Given a distance of about 3,700, the linear diameter of NGC 559 is around 7.5 light years. It is apparently over a billion years old, making it one of the more venerable open clusters in the Milky Way.

NGC 581 (M103)

This open cluster in Cassiopeia contains about 170 stars, the brightest of which are arranged in a neat shape that (with a bit of imagination) resembles a Christmas tree. There is even a bright star at the top of the tree, but unfortunately it isn't actually a member of the cluster and just happens to lie in the same line of sight. In any case, this is one of those objects that looks best through small and medium apertures; through large aperture telescopes the cluster becomes more resolved and more fainter stars are revealed, but this makes the Christmas tree shape more difficult to see. The cluster is quite small, only about 13 arcminutes in diameter, so a medium level of magnification will probably be needed to fully resolve it.

M103 is about 30 light years in diameter and about 8,000 light years away from Earth. M103 is part of a group of open clusters known as the Cassiopeia OB8 Association. All are roughly the same distance from us, and more importantly, seem to be about the same age. They are believed to be clusters that were born from the same star-forming nebula somewhere between 20 and 25 million years ago. There other clusters in Cassiopeia belong to this group and can be easily seen by amateur astronomers: NGC 663, NGC 654, and NGC 659. More is said about each of these clusters elsewhere in this chapter.

M103 is of note in the history of astronomy as being the very last object Charles Messier added to his famous catalog of deep sky objects. Although the list itself runs up to 110 objects, including one duplicate, numbers 104 upwards were added by other people.

NGC 628 (M74)

This is a face-on spiral galaxy in the zodiacal constellation of Pisces. Like most face-on galaxies, its surface brightness is rather low. This means that it is a very difficult object for light polluted skies. To be more precise, while the core itself is fairly bright and easy to spot with apertures upwards of 150 mm under reasonably good suburban skies, you'll probably need quite dark exurban skies to see the spiral arms. It goes without saying you'll want to be fully dark adapted before trying to see this galaxy, and the use of averted vision is helpful, too.

M74 is around 35 million light years away and has an angular diameter of about 10 arcminutes. Its actual diameter would therefore be a little over 100,000 light years. At least two supernovae have been seen in this galaxy. M74 is a member of its own group of galaxies, the M74 Group. The other galaxies in this group are much dimmer and not of any particular interest to amateur astronomers.

M74 is notorious among astronomers who like to do the Messier Marathon. To be seen at all, it needs to be bagged very early on, just after the Sun has set, and seeing galaxies in twilight is a real challenge.

NGC 650 (M76, Little Dumbbell)

The Little Dumbbell in Perseus is a favorite planetary nebula for observers during the autumn and winter months. As its name suggests, it's somewhat like the Dumbbell Nebula (M27, NGC 6853) in Vulpecula, though smaller and less bright. It is still a good nebula for observers under suburban skies, though, and responds particularly well to the use of narrowband and O-III filters. These filters not only improve the contrast of the object, they also make some of the wispy, nebulous regions beyond the boxy core easier to see, especially with averted vision. The star at the center

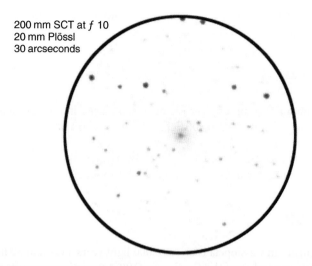

200 mm SCT at ƒ 10
20 mm Plössl
30 arcseconds

Figure 5.7. Like many face-on spiral galaxies, M74 is a difficult object to view under suburban conditions. (Image produced using Starry Night Pro. AllSky data courtesy of Main-Sequence Software Inc.)

of the Little Dumbbell is too faint (magnitude 16.6) to see seen by most amateur astronomer's telescopes.

As is commonly the case with planetary nebulae, estimates of how far away the Little Dumbbell is from Earth vary significantly. The lowest values suggest a distance of around 2,000 light years, while the highest values are well over 10,000 light years. Astronomy books typically give the distance as being around 3,400 light years, but it should be acknowledged that this is by no means a universally accepted value.

Assuming the distance of this object is around 3,400 light years, then the boxy core, which has dimensions about 40 by 90 arcseconds, would be about 0.66 by 1.5 light years.

NGC 654

NGC 654 is an open star cluster in Cassiopeia. It is small in apparent size, about 5 arcminutes in diameter, but quite rich and bright, and looks very nice when viewed with medium to large aperture telescopes under moderate to high levels of magnification.

NGC 654 is about 7,000 light years away and 10 light years in diameter. It also appears much dimmer than it should actually be, implying that it is partially obscured by interstellar dust that lies between us and the star cluster.

NGC 654 is one of four open star clusters in Cassiopeia that are members of the interesting Cassiopeia OB8 Association. The other members are M103, NGC 659, and NGC 663. All these clusters are about 20–25 million years old and were formed in the same nebula.

NGC 659

This small (about 5 arcminutes across) open star cluster in Cassiopeia is quite faint and difficult to see. It is about 6,300 light years away and a little over 9 light years in diameter.

NGC 659 is a member of the Cassiopeia OB8 Association along with M103, NGC 654, and NGC 663. Like them, it is thought to be about 20–25 million years old.

NGC 663

This open cluster in Cassiopeia is about 6,000 light years away and 30 light years in diameter. It is a member of the Cassiopeia OB8 Association, a group of open star clusters that were formed from the same nebulae; other members of this group that can be seen in Cassiopeia are M103, NGC 654, and NGC 659.

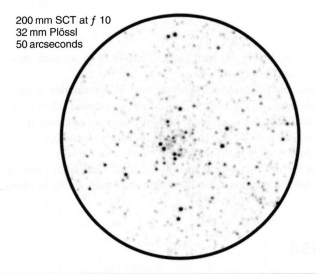

200 mm SCT at ƒ 10
32 mm Plössl
50 arcseconds

Figure 5.8. Cassiopeia contains numerous open clusters; NGC 663 is one of the richest and prettiest. (Image produced using Starry Night Pro. AllSky data courtesy of Main-Sequence Software Inc.)

Through telescopes of all sizes NGC 663 is an attractive target, perhaps one of the best of the NGC clusters in this constellation. It contains no fewer than 100 stars, of which around 15 are bright and easy to see even with small telescopes. Large aperture telescopes will resolve even more of stars. There are a mix of colors on view, with the brightest stars being yellow and many of the background stars blue or white.

NGC 1023

The lenticular galaxy in Perseus is typical of that type, with a morphology in between that of a spiral galaxy and an elliptical galaxy. Through a medium to large aperture telescope it will be seen as an elliptical blur of light with a core that is clearly brighter that the rest of the object, but no obviously discernible arms. Long-exposure photographs show traces of a bar-like structure towards the core, but nothing like as well developed as in a proper barred spiral galaxy.

In general NGC 1023 is a fairly difficult target for suburban astronomers. It isn't particularly bright, and while the core should be visible through large aperture telescopes under reasonably good conditions, the rest of the galaxy is easily overwhelmed by skyglow. It doesn't help that this galaxy has a rather small angular diameter, about 7.5 arcminutes along its long axis. This means that given that only the bright core is easily seen, it can be easily mistaken for a dim, out-of-focus star.

NGC 1023 is about 34 million light years away and a bit over 74,000 light years in diameter. It is the brightest member of a group of galaxies known as the NGC 1023 Group that has its center some 20–21 million years away from us.

NGC 1039 (M34)

M34 is an open cluster in Perseus that can be just about seen with the naked eye under very dark sky conditions. It is an easy object for astronomers using binoculars, and looks nice through telescopes of all apertures.

A medium to large aperture telescope should comfortably resolve between 50 and 80 stars, mostly blue or white, but with a few yellow ones in there as well. However, given that this cluster is about 35 arcminutes in diameter, it is very important to use as low a magnification and as wide a field as you can. Rich-field refractors and fast Newtonian reflectors shouldn't have problems in this regard, but long focal length telescopes, such as SCTs and Maksutov–Cassegrain telescopes, aren't the ideal instruments for looking at big star clusters such as NGC 1039. Try using an f/6.3 reducer-corrector, perhaps in conjunction with a 2-in. wide-field eyepiece like a 35-mm Panoptic.

NGC 1039 is about 1,400 light years away and 14 light years in diameter. It is thought to be of middling age, and while older astronomy books list its age at about 100 million years, more recent estimates suggest an age closer to 190 million years.

NGC 1068 (M77)

This spiral galaxy is the only Messier catalog object in the constellation of Cetus, the fourth-largest constellation in the sky. It is a Seyfert-type galaxy, meaning that it has a bright core that emits a lot of radio waves, a peculiarity taken to imply the presence of a supermassive black hole at the center of the galaxy.

M77 is in fact the brightest Seyfert galaxy, but because of its face-on orientation, its apparent magnitude of 9.6 is a bit misleading. In practice this is a tricky object to see visually, particularly from suburban locations. While its core may well be spotted under good conditions through telescopes 150 mm in aperture or smaller, to see any detail an aperture of at least 200 mm is required. Even with a large aperture telescope the arms are difficult to see, and what will be more apparent instead is a lumpy halo around the bright, almost star-like core.

M77 is believed to be some 47 million light years away and its angular diameter of about 7 arcminutes implies an actual diameter of around 96,000 light years.

NGC 1097

NGC 1097 is a barred spiral galaxy in the southerly constellation of Fornax. It is a member of the Fornax Cluster, one of the richest galaxy clusters nearby to the Local Cluster to which the Milky Way Galaxy belongs. The brightest member is NGC 1316 (see below) but NGC 1097 isn't far off in terms of prominence. Through a telescope with an aperture of 150 mm or more, the bright center of NGC 1097 is easily visible as an almost star-like object. Dark skies and more aperture will reveal the central bar, but the spiral arms are notoriously difficult to see.

NGC 1097 has a small companion galaxy, known as NGC 1097A, that is actually within the range of the NGC 1097's spiral arms. Over time NGC 1097A will presumably be fully absorbed by its larger partner, but in the meantime it is distorting NGC 1097, changing its shape. NGC 1097A is very difficult to see visually, and Stephen

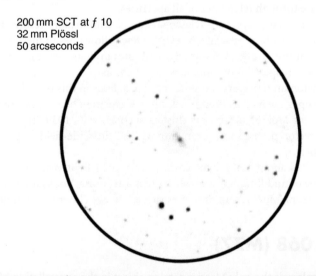

200 mm SCT at ƒ 10
32 mm Plössl
50 arcseconds

Figure 5.9. NGC 1097 is one of the brighter members of the Fornax Cluster of galaxies, but its southerly location means it is difficult or impossible to see from far northern latitudes. (Image produced using Starry Night Pro. AllSky data courtesy of Main-Sequence Software Inc.)

James O'Meara reports seeing this companion galaxy with a 102 mm telescope from a dark sky site, NGC 1097A is more generally considered a challenging target even for astronomers using a 200 mm telescope. Through a telescope all that can be said about NGC 1097A is that it appears as a very dim star-like object. As is always the case with faint but interesting targets: be careful not to imagine seeing something just because you know it's there!

Fornax is a southerly constellation, and far northern observers will find NGC 1097 a difficult galaxy to observe. At the latitude of southern England, it rises less than 7° above the horizon.

NGC 1316 (Fornax A)

Fornax A is a lenticular galaxy in Fornax. Visually, this is the brightest member of the Fornax Cluster, but among professional astronomers it is most notable for being an extremely strong radio source.

At the eyepiece Fornax A looks like a faint elliptical blur about 13 arcminutes in maximum width. It has a bright, almost circular core that should be easy to spot, but the thick elliptical halo around the core is faint and easily obscured by ambient light pollution. Through small telescopes this object can appear almost star-like, but even medium to large aperture telescopes won't show much beyond its broadly elliptical shape.

Fornax A has had a turbulent history. Like other galaxies it contains open star clusters, but astronomers have noted that there are at least two generations of them. Some are filled with hot, young stars, while others contain much older and dimmer stars. The likely explanation is that Fornax A has absorbed smaller galaxies in the past. One such event took place around a 100 million years ago, and the younger star clusters were probably formed by that event. But the older star clusters are the remains of an incident that took place much further back in the past.

Fornax A is around 45 million light years away and 120,000 light years in diameter. About half a degree from Fornax A is NGC 1317, a small spiral galaxy with which it is interacting.

Like other objects in Fornax, this galaxy has a southerly declination and is difficult to see from mid to far northern locations. It is essentially invisible from the latitude of southern England, rising a mere degree or so above the horizon.

NGC 1907

See NGC 1912 (M38) in the showpiece objects section above.

NGC 1952 (M1, Crab Nebula)

The Crab Nebula is probably the most famous supernova remnant in the night sky. Providing your skies are only slightly light polluted, this is an easy object to see. It is,

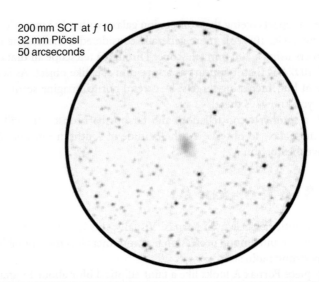

200 mm SCT at *f* 10
32 mm Plössl
50 arcseconds

Figure 5.10. M1 is the supernova most easily viewed from suburban locations. However, although it is fairly bright, a narrowband or O-III filter will be extremely helpful. (Image produced using Starry Night Pro. AllSky data courtesy of Main-Sequence Software Inc.)

however, very small and at low magnifications can be easily overlooked. Despite its popular name, the Crab Nebula isn't particularly crab-like at all, and in fact looks more like a dim ellipse about 6 by 4 arcminutes in size.

There is some debate about whether the Crab Nebula is enhanced by the use of light pollution filters. Under conditions of mild light pollution, a broadband filter will help to make this object stand out against the background sky. Some of its wispier details should be easier to see as well. Because supernova remnants emit light across a broad range of wavelengths, it is often assumed that narrowband and line filters are useless for viewing them. In fact both narrowband and O-III filters can be used to view the Crab Nebula, and while the O-III filter in particular dims the nebula to a considerable degree, it can also make some of the details a bit more contrasting. In short, while some observers prefer the view without a light pollution filter, other observers, the author included, find using filters, at least some of the time, well worth doing.

The Crab Nebula is one of the few deep sky objects for which a precise age can be applied. This is because the Crab Nebula is the remains of a supernova that was first observed by Chinese astronomers in July 1054. At the time, this supernova was incredibly bright, much brighter than Venus, and consequently visible even by day. Given that the Crab Nebula is about 6,500 light years away and it would therefore take 6,500 years for light to reach us from the exploding star, the supernova must have occurred about 7,500 years ago.

The Crab Nebula is about 11 light years in diameter and still expanding. It is an extremely strong source of radio waves and X-rays, and was an important target for astronomers as they developed techniques for studying objects at these wavelengths. At the center of the nebula is a rapidly rotating neutron star of a type known as a pulsar. For more on pulsars, see the entry for NGC 6626 in the chapter on summertime objects.

NGC 2281

NGC 2281 is an open cluster in Auriga. It is large and quite bright, and considered a good target for astronomers using binoculars and wide field refractors and Newtonian reflectors. Given that it has an angular diameter of 14 arcminutes, NGC 2281 is an object that needs to be viewed with a balance of magnification and field of view. For an f/10 200-mm SCT, a 20-mm Plössl is a good eyepiece to use, framing the cluster nicely in the resulting 30 arcminute field of view, but giving a useful ×100 magnification that resolves a good proportion of the stars in the cluster.

Most of the stars in this cluster are blue–white, but there's a yellow supergiant among them. NGC 2281 is about 1,500 light years away and is believed to be around 50 million years old.

NGC 6939 and NGC 6946

NGC 6939 is an open cluster in Cepheus is one of the oldest open clusters that amateur astronomers can readily observe, with an estimated age of around 2 billion

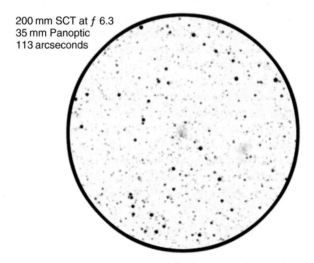

200 mm SCT at f 6.3
35 mm Panoptic
113 arcseconds

Figure 5.11. NGC 6939 is an ancient open cluster that can be placed in the same field of view as the face-on spiral galaxy NGC 6946. (Image produced using Starry Night Pro. AllSky data courtesy of Main-Sequence Software Inc.)

years. It has an angular diameter of less than 8 arcminutes, and at low magnifications and through small aperture telescopes it is difficult to resolve. A medium to large aperture telescope is needed to get good views of this object.

A dim spiral galaxy is nearby, NGC 6946, also known as the Fireworks Galaxy. This object is viewed face-on, and as is often the case with galaxies viewed in this way, only the core is easily seen. Good, dark skies are needed for the much fainter spiral arms to be detected. The distance between us and NGC 6946 is uncertain, but current estimates suggest it is only 10 million light years away, and therefore its angular diameter of a little under 12 arcminutes implies a diameter of only 34,000 light years, making it a relatively small galaxy. Note that some estimates put this galaxy much further away than this, up to 18 million light years, in which case NGC 6946 would have to be much bigger to have the same angular size.

NGC 6939 and NGC 6946 can be viewed at the same time, but for that you'll need a field of view of around a degree since they're about 40 arcminutes apart. If you're using a 200-mm SCT, that won't be possible with 1.25-in. eyepieces unless you use an f/6.3 reducer-corrector as well. If you use a reducer-corrector, a 32-mm Plössl will produce the necessary 1° field of view, and assuming the sky is dark enough, you'll be able to enjoy both the galaxy and the open cluster at the same time.

NGC 6981 (M72)

This is one of two globular clusters in the constellation of Aquarius that made it into the Messier catalog, the other being M2 (NGC 7089). It is a very remote globular clusters some 53,000 light years from Earth, but the fact that it is still relatively easy to see (magnitude 9.2) implies that it must be an inherently extremely bright object. But having said that, M72 is widely considered one of the less impressive Messier globular clusters, and its relatively southerly location means that far northern observers will find it tricky to observe.

Like all globular clusters, M72 benefits from both aperture and magnification. It isn't a good target for small to medium aperture telescopes, but a 200-mm SCT is big enough to make it possible to resolve some of the outer stars. M72 is rated as a Class IX object on the Shapley–Sawyer scale M72, implying that it is a relatively loose globular cluster. This is true enough, but given its small angular size, about 6 arcminutes, and its overall faintness, resolving this globular cluster to the core is impossible without using an extremely large telescope.

NGC 7089 (M2)

M2 is by far the brighter of the two globular clusters in Aquarius and an excellent target for telescopes of all sizes. It is easily visible with binoculars, and small and medium aperture telescopes will reveal a granular patch of light about 16 arcminutes in width.

M2 is a very condensed globular cluster, rated at Class II on the Shapley–Sawyer scale. It is also very large, with an actual diameter of over 170 light years, making about the same size as the Great Globular Cluster in Hercules. But whereas that globular cluster is only 25,000 light years away, M2 is more than 37,000 light years away, and therefore appears smaller and fainter. Like other globular clusters this is a very ancient object: M2 has an estimated age of about 13 billion years.

Like most other globular clusters, M2 is orbiting the galactic core. However, it does so along a very eccentric orbit. At its closest, M2 is within 23,500 light years of the center of the Milky Way Galaxy, but at its furthest it is some 171,000 light years out! To put this in perspective, consider that the Large Magellanic Cloud orbits the Milky Way Galaxy at a distance of about 160,000 light years from the core.

NGC 7331

NGC 7331 is a spiral galaxy in Pegasus. It is of medium size (angular diameter about 10 arcminutes) but not very bright, though its core at least should be readily visible under exurban and possibly suburban conditions. It is viewed at an angle and under good conditions looks like a thin ellipse.

This galaxy is remarkable because its core rotates in the opposite direction to the disk (i.e., the spiral arms). Where the core and the disk meet, gravitational interactions cause gas and dust to collide, triggering star formation.

NGC 7331 is also considered noteworthy because its size and mass seem to be very similar to that of our Milky Way Galaxy, to the degree that the two galaxies are

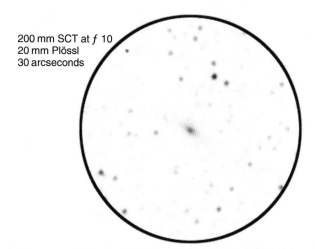

200 mm SCT at ƒ 10
20 mm Plössl
30 arcseconds

Figure 5.12. NGC 7331 is a spiral galaxy viewed edge on; although quite faint, its core at least is detectable under suburban skies. (Image produced using Starry Night Pro. AllSky data courtesy of Main-Sequence Software Inc.)

sometimes described as 'twins.' It should be noted, however, that NGC 7331 does not belong the Local Group of galaxies that includes the Milky Way but is a member of its own group of galaxies, known as the NGC 7331 Group.

NGC 7331 is about 40 million light years away.

NGC 7789 (Caroline's Rose)

NGC 7789 is an attractive open cluster in Cassiopeia that looks good through tele-scopes of all sizes. It was discovered by Caroline Herschel in 1783 and has sometimes been known as Caroline's Rose. Small aperture telescopes reveal a small, round, almost nebula-like patch of light, but medium to large apertures will resolve many more stars packed in an area of space about 16 arcminutes in diameter. There are in fact over 1,000 stars in this cluster. It is also remarkably circular, and with a bit of imagination the whorls of stars can be likened to the petals of a rose.

This is a fairly old cluster, with an estimated age of 1.6 billion years. Besides the usual blue and white stars, there are also a few yellow and orange stars; these are red giants that have used up the reserves of hydrogen they need for normal nuclear reactions and are rapidly approaching death.

NGC 7789 is about 7,600 light years away and about 35 light years in diameter.

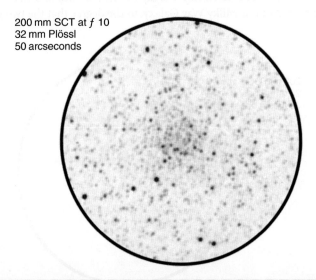

200 mm SCT at ƒ 10
32 mm Plössl
50 arcseconds

Figure 5.13. NGC 7789 is a fairly densely packed open cluster that has an almost nebula-like quality at low to medium powers. (Image produced using Starry Night Pro. AllSky data courtesy of Main-Sequence Software Inc.)

Obscure and Challenging Deep Sky Objects

NGC 40

NGC 40 is a planetary nebula in Cepheus. It is reasonably bright and easy to see with medium to large aperture telescopes. Under suburban conditions at least the nebula appears to be disc-shaped with no obvious details or variations in brightness across its surface. A narrowband filter is helpful, improving contrast and making it easier to see the nebula's shape.

NGC 40 has a diameter of a little over half a degree, so substantial levels of magnification will be required to see this object properly, upwards of ×150. It is about 3,500 light years away and 36 light years in diameter.

NGC 129

This open cluster in Cassiopeia is large but loose, and perhaps looks rather better through binoculars and wide field telescopes than it does through the average SCT or Maksutov–Cassegrain telescope. It is certainly a good idea to use a low magnification when observing this object, ideally with an f/6.3 reducer-corrector as well to widen the field of view as much as possible. With an angular diameter of over 21 arcminutes, an f/10 SCT equipped with a 32-mm Plössl works reasonably well, providing enough of a background that the sparse cluster that is NGC 129 stands out clearly.

NGC 129 contains several giant stars including a Cepheid variable known as DL Cassiopeiae (SAO 21446). This variable star has a period of almost exactly 8 days and fluctuates in brightness from 8.7 down to 9.28. Cepheid variables are important because of the known relationship between their intrinsic brightness and the periodicity. By measuring their periodicity, in this case 8 days, you can work out their intrinsic brightness, which in the case of DL Cassiopeiae is (at its maximum) a magnitude of –4.2. Since the apparent brightness of a star (i.e., how bright it seems to us) is a function of how far away it is, if you know its intrinsic brightness and apparent brightness, you can calculate its distance.

DL Cassiopeiae is in fact one of the best studied Cepheid variables in the sky, and because it is a known member of NGC 129, it can be used to accurately calculate how far away the cluster is from Earth. Recent estimates place DL Cassiopeiae at a distance of about 6,600 light years, quite a bit higher than the 5,400 light year value widely assumed in the past. If the higher number is accurate, then NGC 129 must have a linear diameter of around 33 light years.

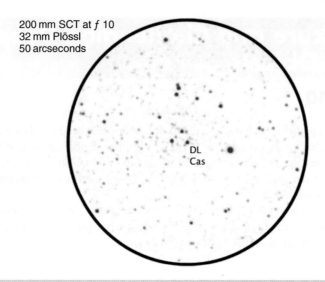

200 mm SCT at f 10
32 mm Plössl
50 arcseconds

DL
Cas

Figure 5.14. NGC 129 is a loose, fairly nondescript open cluster notable for containing a Cepheid variable star, DL Cassiopeiae, that allows the distance of this open cluster to be estimated accurately. DL Cassiopeiae is shown here at the exact center of the field of view. (Image produced using Starry Night Pro. AllSky data courtesy of Main-Sequence Software Inc.)

NGC 188

This open star cluster in Cepheus is less than 4° from Polaris. It is a rich cluster that contains more than 200 stars, but with an angular size of 14 arcminutes it cannot be resolved through binoculars or small telescopes. Medium to large aperture telescopes present an enchanting sight, and while none of its constituent stars are bright, they are densely packed, and the overall impression is almost like a globular cluster that's been squeezed and has burst.

Stephen James O'Meara has pointed out that NGC 188 is probably not well known to amateur astronomers only because it is in a part of the sky difficult to view through equatorially-mounted telescopes. Users of go-to telescopes on alt-azimuth mounts won't have to deal with any such problems.

NGC 188 is a very old open cluster, likely some 5 billion years old. As such, it is among the oldest open clusters in the Milky Way Galaxy. Current estimates place the distance between Earth and NGC 188 at around 5,000 light years. Given that, its 14 arcminute apparent diameter would imply an actual diameter of a little more than 20 light years.

NGC 189

NGC 189 is an open cluster in Cassiopeia only 3.7 arcminutes in diameter. Because it is so small, this object is best viewed using medium to large aperture

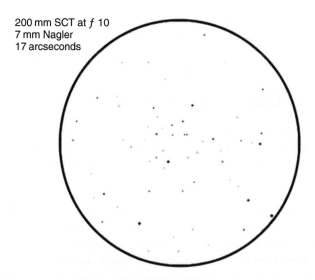

200 mm SCT at ƒ 10
7 mm Nagler
17 arcseconds

Figure 5.15. NGC 189 has a very small angular diameter and requires both ample aperture and significant levels of magnification to be resolved. (Image produced using Starry Night Pro. AllSky data courtesy of Main-Sequence Software Inc.)

telescopes. This isn't a particularly impressive star cluster, and moderately high levels of magnification will be needed to resolve it completely. Like many such clusters in Cassiopeia, it isn't easy discerning which stars belong to the cluster and which stars are simply in the background field.

NGC 189 is about 3,500 light years away and is believed to be about 20 million years old. This cluster is of slight historical interest because of confusion over its identity. Caroline Herschel discovered the cluster in 1783, but her brother William Herschel mistakenly identified her description with the cluster we now call NGC 381. His son John Herschel came across the cluster in 1829 and for a long time it was assumed to be his discovery. It wasn't until 2005 that historians recognized that NGC 189 had in fact been first discovered by Caroline Herschel.

NGC 246

The planetary nebula NGC 246 is a small, dim planetary nebula in the constellation of Cetus.

NGC 246 is a challenging target for owners of telescopes less than 200 mm in aperture, particularly under suburban conditions. Several stars lie in between us and the nebula, and the light from these can make it difficult to see the nebula. Like many planetary nebulae, views of NGC 246 are improved through the use of a narrowband or O-III filter, and in this case the fact the foreground stars are dimmed as well is an added (and welcome) bonus.

NGC 246 has an angular width of about 4 arcminutes, and given that it some 2,000 light years away from us, this implies the nebula is in fact a bit over 2 light years across.

NGC 281 (Pac-Man Nebula)

Strictly speaking NGC 281 is an emission nebula in the constellation of Cassiopeia, but under suburban conditions only the open cluster known as IC 1590 is obvious. The use of a narrowband or O-III filter may help make the surrounding nebula easier to see under exurban conditions, but this is really an object that looks best under dark skies.

NGC 281 is a star-forming nebula, and the bright open cluster IC 1590 is just one of several such groups of newborn stars. The nebula is about 9,400 light years away. It gets its popular name from its shape; in long-exposure photographs especially, this nebula does indeed resemble the popular video game character.

NGC 584

NGC 584 is an elliptical galaxy in the constellation Cetus. It isn't a difficult galaxy to spot since it has a very bright core, but actually seeing its overall shape is much harder. You need good dark skies for the spindle-like halo around the core to become apparent. As with any other galaxy, using a broadband filter may help slightly, but otherwise this is a tricky object for suburban observers.

NGC 584 is about 78 million light years away and some 80,000 light years in diameter. It is a member of its own group of galaxies, the NGC 584 Group within the Cetus-Aries Cloud.

NGC 598 (M33, Triangulum Galaxy)

The Triangulum Galaxy is a spiral galaxy in the constellation of Triangulum that is notorious for being large but difficult to see. The problem is that its size and face-on orientation spreads out widely the light it emits, resulting in very low overall contrast compared with the background sky. Its size is remarkable, though, and this is one reason it is a favorite target for astrophotographers. After the Andromeda Galaxy, it has the greatest apparent size of any galaxy in the northern sky, being about 66 by 38 arcminutes in size. This means it takes up more of the sky than the full Moon!

Under pitch black skies the Triangulum Galaxy is just visible to the naked eye, and is an easy object for binoculars and wide field telescopes. However, it has a very low surface brightness, and unlike many other galaxies, the core isn't substantially brighter than the arms. Skyglow can make the galaxy difficult to see even away from artificial sources of light pollution, and a broadband filter is therefore well worth using.

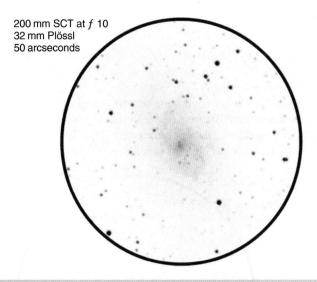

200 mm SCT at ƒ 10
32 mm Plössl
50 arcseconds

Figure 5.16. The Triangulum Galaxy has a very low surface brightness, and even its core is difficult to detect under suburban conditions, let alone its spiral arms. (Image produced using Starry Night Pro. AllSky data courtesy of Main-Sequence Software Inc.)

Astronomers under suburban skies will find this object a severe challenge. Given its size, a wide field of view is essential, and users of f/10 200-mm SCTs should use a reducer-corrector to widen the field of view offered by eyepieces such as a 32-mm Plössl.

The Triangulum Galaxy has an actual diameter of about 60,000 light years, making it smaller than both the Milky Way and the Andromeda Galaxy. Nonetheless, it is still a large galaxy in general terms, and with our own Milky Way Galaxy and the Andromeda Galaxy, is one of the biggest members of the Local Group of galaxies.

At about 3 million light years away, the Triangulum Galaxy is a little further away from us than the Andromeda Galaxy, which is about 2.5 million light years away. As such, it's probably the most distant object visible to the naked eye.

NGC 772

The constellation of Aries contains few deep sky objects of interest to amateur astronomers, and while NGC 772 is probably the best of them, it's still a dim and difficult target. It is a spiral galaxy about 130 million light years away, viewed approximately face-on. A large aperture telescope is needed to see NGC 772, and even then only a small, dim, and hazy patch of light will be visible.

NGC 908

NGC 908 is a spiral galaxy in Cetus. It is notable for being one of what are called starburst galaxies, galaxies that have areas of intense star formation dotted around their arms. Typically these are triggered by interactions with other galaxies that moved too close. These would have to occurred quite recently, since star, otherwise the galaxy would have settled down and the areas of star formation dissipated. It's assumed that NGC 908 drifted past another galaxy at some point in the past, but which galaxy isn't known because there aren't any obvious candidates nearby.

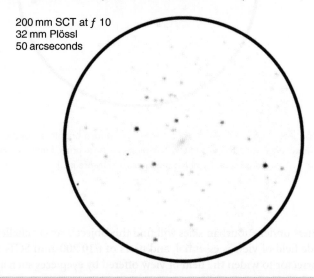

200 mm SCT at *f* 10
32 mm Plössl
50 arcseconds

Figure 5.17. NGC 908 is a dim spiral galaxy that is not easy to see under suburban conditions. (Image produced using Starry Night Pro. AllSky data courtesy of Main-Sequence Software Inc.)

Through a large aperture telescope this is a fairly typical small, faint, elliptical smear of light. Its core is noticeably brighter than the halo, but otherwise no detail is obvious. It is best observed under exurban or dark sky conditions, but the core at least can be glimpsed from suburbs assuming the ambient light pollution isn't too severe. NGC 908 is about 65 million light years away and measures about 75,000 light years in diameter.

NGC 936

NGC 936 is a dim barred spiral galaxy in Cetus. It has a reasonably accessible core that can be spotted with medium aperture telescopes, but even through large apertures the outer halo will prove elusive. Primarily an object for observation under dark sky conditions, but worth a shot from good suburban and exurban locations.

NGC 936 is about 54 million light years away. Its angular diameter of 5.2 arcminutes corresponds to an actual diameter of around 82,000 light years.

NGC 1275 (Perseus A)

NGC 1275 is a lenticular galaxy in Perseus and an example of the Seyfert class of galaxy, a type that emits massive amounts of radiation from the core suggestive of an accretion disk surrounding a supermassive black hole. When viewed in the X-ray spectrum, NGC 1275 is in fact one of the brightest objects in the sky.

NGC 1275 is about 250 million light years away, so although it is very large in absolute terms, almost 190,000 light years across, its apparent size is tiny, about 2.6 arcminutes. It is difficult to see through medium aperture telescopes, and a large aperture telescope will probably be required under suburban conditions. At best, this object looks like a dim, small, and fuzzy patch of light.

NGC 1275 is the brightest member of the Perseus Cluster of galaxies. There are in fact thousands of galaxies in this cluster, with NGC 1275 the giant in the center, pulling in and absorbing any galaxies that get too close. Astronomers with medium to large aperture telescopes can spend some time exploring the area around NGC 1275 looking for some of the other members of the Perseus Cluster such as NGC 1272 and NGC 1278.

NGC 1342

This open cluster in Perseus is loose and covers a full 14 arcminutes of space, and unsurprisingly looks best at relatively low magnifications. It can be spotted through telescopes of all sizes, including binoculars, but medium to large apertures will be needed to resolve the cluster fully.

NGC 1342 is about 2,200 light years away and around 400 million years old.

NGC 1360

NGC 1360 is a planetary nebula in the southerly constellation of Fornax. Like other objects in Fornax, it is not an easy target for far northern observers. At the latitude of Southern England it only rises 12° or so above the horizon. But that aside, this is quite a bright nebula well within the range of medium the large aperture telescopes. It is rather diffuse though, spanning some 6.5 arcminutes in diameter, so while bright, its light is spread out, and consequently it suffers from a lack of contrast when compared to the background sky. However, NGC 1360 will be easier to see if a narrowband or O-III filter is used; these get rid of any skyglow and help to make the nebula easier to detect.

NGC 1360 is about 850 light years away and a little over 1.6 light years in diameter. At its center is a magnitude 11.4 star that should be detectable through medium to large aperture telescopes. Under good conditions the nebula appears to be mottled rather than uniform in texture. As is often the case with planetary nebulae, increasing magnification is very helpful; not only does this make details larger and therefore easier to discern, it also dims the background sky, heightening overall contrast.

NGC 1365

NGC 1365 is a barred spiral galaxy in Fornax. It is of moderate brightness, and the core at least should be visible in telescopes of medium to large aperture. Under dark sky conditions large aperture telescopes show not just the core but also the bright bar that runs across the core and hints of its two wispy arms. More realistically, amateur astronomers observing under suburban conditions will likely have to content themselves with the fairly bright elliptical core.

As a barred spiral galaxy with just two major arms, NGC 1365 has acquired a certain degree of celebrity in recent years as a possible analog for our own Milky Way Galaxy. Until recently it has always been assumed that the Milky Way Galaxy has four major arms, but some studies now suggest that only two of its arms, the Perseus Arm and the Scutum-Crux (or Centaurus) Arm qualify as major arms, with all the others being minor arms. If that's the case, then it probably looks and functions a lot like NGC 1365.

NGC 1365 is about 60 million light years away and nearly 200,000 light years across, making it one of what are sometimes called supergiant galaxies.

NGC 1491

NGC 1491 is an emission nebula in Perseus. It is small (only about 3 arcminutes across) and difficult to see except under quite dark conditions. It is best viewed at moderate levels of magnification ($\times 50$ upwards) and with a narrowband filter. Large aperture telescopes will be required to see NGC 1491.

NGC 1491 is 12,000 light years from Earth and about 10 light years in diameter.

NGC 1528

This large, bright open cluster in Perseus is a particularly nice object when viewed at low magnifications. With an angular diameter of over 23 arcminutes, a wide field is essential. In the case of a standard 200-mm SCT, a 32-mm Plössl is a good eyepiece to start with, producing a big field that frames the cluster reasonably well.

As is always the case with these large open clusters, expanding the field of view (for example by using an f/6.3 reducer-corrector) does much to improve the overall impression that the object makes, but increasing the magnification can be useful for teasing out any unresolved members of the cluster.

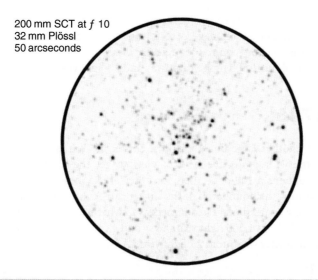

200 mm SCT at f 10
32 mm Plössl
50 arcseconds

Figure 5.18. Given its large angular size, NGC 1528 is best observed in a low magnification eyepiece. (Image produced using Starry Night Pro. AllSky data courtesy of Main-Sequence Software Inc.)

NGC 1528 is about 2,400 light years away and approximately 16 light years in diameter. The cluster contains at least 50 stars.

NGC 1582

One of several large but star-poor open clusters in Perseus, NGC 1582 is an attractive object visible through telescopes of all sizes. It is primarily an object for binoculars and wide-field telescopes, and measures some 37 arcminutes in diameter. For best effect, use a low power, wide-field eyepiece. A 32-mm Plössl when used with a 200-mm SCT just about fits the whole cluster in the resulting 50 arcminute field of view, but using a reducer-corrector as well helps significantly. As is often the case with these loose, scattered clusters, the more background sky you can bring into the view, the more obvious (and impressive) the cluster becomes.

NGC 1582 is about 3,600 light years distant and 38 light years in diameter. Its age is thought to be around 300 million years.

NGC 1647

NGC 1647 is a large and rather scattered open cluster in Taurus. It spans a good three-quarters of a degree, and though it fits into the field of view offered by a 32-mm Plössl when used with a standard f/10 SCT, it won't make much of an impression.

This cluster is best appreciated through rich-field, wide-angle telescopes, but if you must use an SCT, an f/6.3 reducer-corrector and a wide field 2-in. piece (such as a 35-mm Panoptic) will be required for a decent view.

NGC 1647 is about 1,800 light years away and 23 light years in linear diameter.

NGC 1746

NGC 1746 is another large but poor open cluster in Taurus. As with NGC 1647, you'll need the widest possible field of view to get the most from this cluster. Although it contains a 100 or so members, they are so loosely scattered across the sky that discerning the boundaries of the cluster can be difficult. In fact there's some argument among professional astronomers as to whether this is a true cluster at all.

This cluster is about 1,500 light years away and 42 arcminutes in diameter, implying an actual diameter of around 23 light years.

NGC 1807 and NGC 1817

The center points of these two open clusters are only 24 arcminutes apart, and it isn't difficult to get them into the same field of view. For a 200-mm SCT, a 32-mm Plössl will do the job, but using a reducer-corrector to widen the field of view a bit more, producing a nicer overall impression of the two clusters set against the background sky.

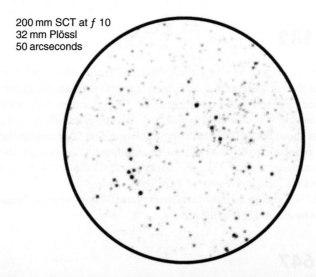

200 mm SCT at ƒ 10
32 mm Plössl
50 arcseconds

Figure 5.19. NGC 1807 (bottom left) and NGC 1917 (top right) look their best when pulled into the same field of view. (Image produced using Starry Night Pro. AllSky data courtesy of Main-Sequence Software Inc.)

Of the two, NGC 1807 contains the brighter stars but is somewhat less dense than NGC 1817, so there's an interesting contrast to be drawn by comparing them. In terms of size, there's not much to choose between them, NGC 1807 being 17 arcminutes in width, and NGC 1817 only marginally smaller, at 16 arcminutes across.

On paper at least, NGC 1807 is about 5,900 light years away and 29 light years in diameter. It is not altogether clear that this cluster really is an open cluster at all. Things are more secure when it comes to NGC 1817, which is a true cluster some 6,400 light years away and 30 light years in diameter.

NGC 1893

NGC 1893 is a young open cluster in Auriga situated some 12,000 light years away. It is particularly notable because it contains a faint emission nebula, IC 410, more than 100 light years in diameter. This nebula is very faint, and suburban astronomers probably won't be able to see it, even with the largest go-to telescopes. Under dark sky conditions the nebula can be detected with telescopes from 300 mm upwards, particularly if equipped with a narrowband or O-III filter.

Returning to the open star cluster, current estimates indicate an age of just 4 million years. That's far younger than even the Pleiades, which are thought to be about 100 million years old. The nebula that surrounds this cluster is the cloud of gas and dust from which these stars were born.

NGC 1893 has an angular diameter of 11 arcminutes, and moderately high magnification is required to resolve all its members.

NGC 1931

NGC 1931 in Auriga is an open star cluster in a combination emission and reflection nebula that measures just 3 arcminutes in diameter. It is quite distant, perhaps as much as 10,000 light years away, and consequently rather faint. Coupled with its small apparent size, NGC 1931 is a tricky target for observers using small aperture telescopes. Although the star cluster will be detectable, resolving the stars that make up the cluster will be hard, and the object will look more like a blurry patch of light. The nebula will not be visible.

Large aperture telescopes resolve more stars, particularly at medium to high magnifications, but the nebula remains hard to see unless the sky is dark. If a narrowband or O-III filter is used the emission nebula part of the object will become easier to see, but such filters will dim the reflection nebula. A broadband filter might be used in slightly light polluted skies, but such a filter is unlikely to help if the light pollution is more serious than this.

NGC 1931 is rather similar to the Orion Nebula (NGC 1976, M42) in basic type, being the birthplace of new stars. Like the famous Trapezium Cluster in the Orion Nebula, the small cluster of stars visible in NGC 1931 are young stars emerging from the gas and dust that made them. As they do, they emit radiation that causes the nebula to glow.

NGC 6946

The constellation of Cygnus is not often associated with galaxies, lying as it does across one of the richest parts of the Milky Way. But on its border with Cepheus there lies a very attractive if dim spiral galaxy, NGC 6946. Although difficult, probably impossible to see under suburban conditions, observers under reasonably dark exurban skies using 200 mm apertures shouldn't have too much trouble seeing the bright core of this object. But as is often the case with galaxies, particularly ones oriented more or less face-on to us, the wispier halo around the core is difficult to see except under dark sky conditions.

By galactic standards NGC 6946 is quite nearby, only about 18 million light years away, and it has sometimes been considered a member of the Local Group of galaxies. This is not now believed to be the case, and it is thought to be a member of the Coma-Sculptor Cloud instead. NGC 6946 has an angular diameter of just over 11 arcminutes, implying a linear diameter of around 58,000 light years.

NGC 6994 (M73)

M73 is an asterism of four stars in the constellation of Aquarius. Initially it was considered a very sparse open cluster, and that's presumably what Charles Messier thought it was when he added it to his famous catalog. He even described a certain amount of nebulosity around the object, though subsequent astronomers have found no trace of this during their own observations.

In recent years there's been debate among astronomers over the nature of M73. On the one hand some argue that it is merely a chance alignment of stars, while others argue that while sparse, these stars actually are the last remaining members of an open cluster that otherwise drifted apart long ago. Supporters of the open cluster theory point maintain that the odds of four, equally bright stars being so close together are, to coin a phrase, astronomical. As such, M73 is an informative example of what happens when the gravitational forces between open star cluster members are no longer sufficient to keep them bound together.

Nonetheless, the weight of evidence seems to imply M73 isn't an open cluster at all. The four stars may seem to be in the same place from our perspective, but in fact they're lying at a range of different distances and traveling in different directions.

NGC 7023 (Iris Nebula)

NGC 7023 is a faint reflection nebula in Cygnus that surrounds a cluster of moderately bright stars. The cluster itself is easy to see, with the brightest star in the cluster,

SAO 19158, shining at a respectable magnitude 7.2. The reflection nebula is more of a challenge, even though by the standards of reflection nebulae, NGC 7023 is one of the easiest such objects to observe. A broadband light pollution filter is especially useful if ambient light pollution is minimal, and averted vision will also help.

As with other reflection nebulae, NGC 7023 contains particles of dust that reflect light from a nearby star, in this case SAO 19158. The nebula is about 1,400 light years away from us and has an angular diameter of about 10 arcminutes, implying an actual width of about 4 light years.

NGC 7129

NGC 7129 in Cepheus is a very young star cluster with an associated reflection nebula. As is often the case with objects of this type, the nebula part is difficult to see under suburban conditions. A broadband filter can help a little, but moderate to high levels of light pollution will render such accessories ineffective. Narrowband and line filters are of no use at all. In most cases, at least reasonably dark exurban conditions will be needed to see the nebula, and even then, averted vision will usually be required.

On the other hand, at least a few of the stars that make up the cluster are bright and easy to see, and resolving around half a dozen stars is well within the grasp of large aperture telescopes. Although unimpressive when observed visually under suburban conditions, NGC 7129 is noteworthy. It is one of the youngest open clusters in the sky, with around 130 members that are less than a million years old.

The nebula is 19 arcminutes in width, though the brighter stars form a much smaller clump within this area. It is about 3,000 light years away.

NGC 7160

NGC 7160 is a small open cluster in Cepheus. It is only about 7 arcminutes in angular diameter, and so it needs to be viewed under reasonably high levels of magnification to be adequately resolved. Two bright (magnitude 7 and 8) stars will be immediately spotted and obvious even through medium aperture telescopes, but there are three magnitude 9 stars arranged in a small triangle that should be visible as well.

NGC 7160 is about 2,500 light years away and is roughly 18 million years old.

NGC 7235

This open star cluster in Cepheus is small, 4 arcminutes in width, and needs to be viewed at medium to high magnifications. Medium to large aperture telescopes will reveal about a dozen stars, the brightest of which shines at magnitude 8.8. NGC 7235 is about 13,000 light years away and measures more than 15 light years in diameter.

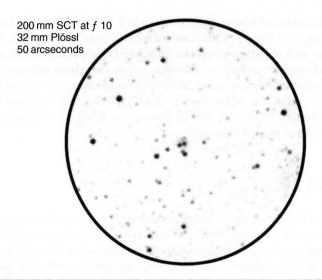

200 mm SCT at ƒ 10
32 mm Plössl
50 arcseconds

Figure 5.20. The open star cluster part of NGC 1529 is quite easy to see, but the nebula will be a challenge under suburban conditions. Because it is a reflection nebula, narrowband and line filters aren't of any value at all, but a broadband filter might offer a little benefit. (Image produced using Starry Night Pro. AllSky data courtesy of Main-Sequence Software Inc.)

NGC 7606

NGC 7606 in Aquarius is a spiral galaxy about 97 million light years away. It has a fairly bright core, but although viewed almost edge-on, it is still quite a faint galaxy and notoriously difficult to see from suburban locations.

A large aperture telescope and averted vision are essential, and even then little more will be seen other than its basic shape. Far northern observers will find things complicated by the fact NGC 7606 doesn't rise very far above the southern horizon. From the point of view of an observer in southern England, it only gets to about 30° above the horizon.

NGC 7635 (Bubble Nebula)

See NGC 7634 (M52) in the showpiece objects section above.

NGC 7793

NGC 7793 in Sculptor is one of the brighter galaxies in the Sculptor Group, but its southerly location makes it a difficult target for northern hemisphere observers. At the latitude of southern England, it barely reaches 6° above the horizon, and

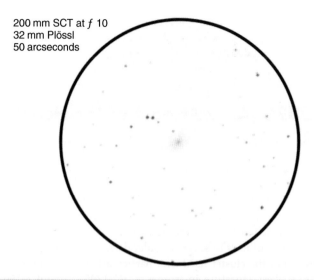

200 mm SCT at ƒ 10
32 mm Plössl
50 arcseconds

Figure 5.21. NGC 7793 is a fairly bright galaxy, but its southerly location means far northern observers will have trouble viewing this object. (Image produced using Starry Night Pro. AllSky data courtesy of Main-Sequence Software Inc.)

between ambient light pollution, dust in the atmosphere, and obstructions like trees and buildings, this makes it essentially impossible for suburban observers to see at all. Going further south will help, but observers in the American Midwest will still find NGC 7793 a mere 16° or so above the horizon.

From the observers point of view, NGC 7793 is in many ways similar to the Triangulum Galaxy (M33, NGC 598). It has the same basic shape, being an almost face-on spiral galaxy, but like the Triangulum Galaxy it also has a notoriously low surface brightness, meaning that it can be very difficult to see except under dark sky conditions. A broadband filter will cut out natural skyglow and a certain amount of light pollution, but NGC 7793 remains a tough target for suburban observers though observers under reasonably dark exurban conditions should be able to spot its bright core.

NGC 7793 is about 27,000 light years away. It has a maximum angular width of about 9 arcminutes, implying a linear diameter of around 70,000 light years.

Colorful and Curious Stars

SAO 4479 (Collinder 463)

SAO 4479 is one of around 30 stars in an open cluster known as Collinder 463. Located in Cassiopeia, this cluster is loose and spread out across 40-odd arcminutes of space, so a low power, wide-field eyepiece is essential. SAO 4479 is the brightest star in the group, at about magnitude 8.2, and this cluster looks its best only when large aperture telescopes are used, though it can be seen easily enough through small and medium aperture telescopes.

SAO 12326 (At the Heart of the Heart Nebula)

This star is in the center of a loose open cluster with associated nebulosity in Cassiopeia referred to as IC 1805 by professional astronomers but better known among amateur astronomers as the Heart Nebula. With an angular diameter of about a degree, a low power, wide-field eyepiece is required to see this very young cluster in its entirety. The nebulosity spans more than 2.5° and has a notoriously low surface brightness. It is difficult to see even under dark sky conditions, let alone suburban conditions, but the cluster itself isn't particularly challenging for astronomers with medium to large aperture instruments. IC 1805 is about 6,150 light years away.

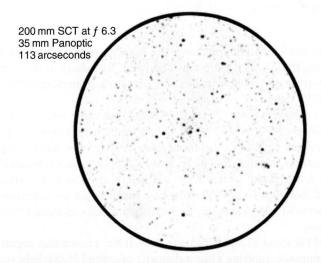

200 mm SCT at ƒ 6.3
35 mm Panoptic
113 arcseconds

Figure 5.22. SAO 12326 is the brightest star at the center of the loose open cluster IC 1805. (Image produced using Starry Night Pro. AllSky data courtesy of Main-Sequence Software Inc.)

SAO 21446 (DL Cassiopeiae)

See NGC 129 in the obscure and challenging deep sky objects section.

SAO 21732 (Eta Cassiopeiae)

Eta Cassiopeiae consists of no fewer than eight components, but the two of interest to astronomers are the yellow magnitude 3.5 primary star and its reddish magnitude 7.5 companion. They are easily resolved through small aperture telescopes, being separated by almost 13 arcseconds. Part of their charm comes from the color contrast between the two stars, with some observers finding that the fainter of the two stars seems purple rather than red or orange. The primary is very Sun-like in its size and composition, and given it has a companion star as well, it makes an interesting northern sky counterpart to the southern sky Alpha Centauri, another Sun-like star.

SAO 23100 (Stock 2)

SAO 23100 is one of dozens of eight to tenth magnitude stars visible in the open cluster known as Stock 2. Located on the very edge of Cassiopeia, the whole cluster spans about a degree, but it is famous for being overlooked by amateurs given its close proximity to the Perseus Double Cluster. It's worth making a quick detour to, though, especially if your telescope is equipped with an f/6.3 reducer-corrector and a 2-in. eyepiece such as a 35-mm Panoptic. Stock 2 is a little over 1,000 light years away and more than 17 light years in diameter.

SAO 23469 (Collinder 29)

Collinder 29 is a moderately rich cluster in Perseus that contains 30 stars scattered across an area a little under half a degree in diameter. It is only a couple of degrees away from the Perseus Double Cluster, and all three clusters can be placed in the same field of view when wide-field telescopes are used. You will need a field of view of about 4° for this to be possible, which means that it isn't possible with the standard SCTs and Maksutov–Cassegrain telescopes.

SAO 23655 (Eta Persei)

Eta Persei is a pretty double star with an orange magnitude 3.8 primary and a blue magnitude 8.5 secondary. They are quite widely separated, a little over 28 arcseconds apart, and so the significant difference in brightness is much less of a hurdle than it might otherwise be. Having said that, Eta Persei is a tricky split for small aperture telescopes, and a good test of both seeing conditions and optical quality (not to mention observing skill!).

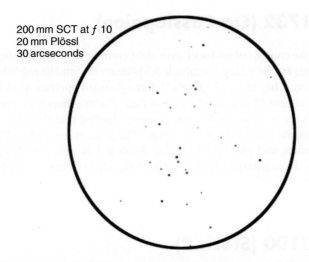

Figure 5.23. SAO 12326 is the brightest member of the open star cluster known as Collinder 29. (Image produced using Starry Night Pro. AllSky data courtesy of Main-Sequence Software Inc.)

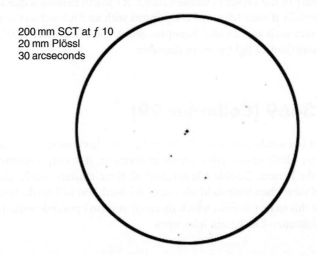

Figure 5.24. Eta Persei is a nice double star easily split. (Image produced using Starry Night Pro. AllSky data courtesy of Main-Sequence Software Inc.)

SAO 33693 (Mu Cephei, Herschel's Garnet Star)

The Garnet Star is one of the prettiest stars in the sky. Small to medium aperture telescopes reveal a rich red star. Large aperture telescopes tend to make the star so bright that its coloration is less obvious.

From a scientific point of view this star is notable as well. It is the largest star known, a huge supergiant star with a diameter 2,400 times that of the Sun (or 22 times the distance between the Sun and Earth). It is also a long period variable, varying from magnitude 3.4 to 5.1 across a period of several years. Its variation isn't regular, though, with its periodicity ranging from about 740 up to 4,400 days.

The Garnet Star has something like 250,000 times the intrinsic luminosity of the Sun, and the fact that it is a comparatively dim star when seen from our place in the galaxy implies that it is also a tremendously distant star as well. Estimates range from 1,630 up to 5,200 light years.

SAO 34508 (Delta Cephei)

A good case can be made for saying that Delta Cephei is one of the most important stars in the night sky. It was identified as a variable star by the English astronomer John Goodricke, varying between magnitudes 3.5 and 4.4 in a period of 5 days 8 hours and 48 minutes. It is conveniently located near to two other stars of constant brightness, so estimating the brightness of Delta Cephei at any given time isn't too difficult. These stars are magnitude 3.4 Zeta Cephei (SAO 34137) and magnitude 4.2 Epsilon Cephei (SAO 34227).

Delta Cephei varies in a very distinctive way, so that a graph of its brightness against time looks like the teeth on a saw, rising in brightness very quickly and then taking much longer to dim back down again. Not long after its discovery, a friend of Goodricke's, Edward Pigott, noticed that another star, Eta Aquilae, varied in the same way. Eventually thousands of such stars were identified, and collectively they have become known as Cepheid variables.

Cepheid variables are bright stars because they expand and contract. This is different from, say, a Mira-type variable that varies in brightness because of the orbit of an unseen companion star. When a Cepheid variable swells up, its surface cools down, so it emits less light. It quickly reaches its maximum size, at which point pressure in the core drops below a certain threshold value and the star slowly shrinks back down again. Once it reaches its minimum size the relatively cool atoms become ionized again, and the star begins swelling up once more.

In 1908 the American astronomer Henrietta Swan Leavitt described one aspect of Cepheid variables that revolutionized astronomy: Cepheid variables exhibit a tight relationship between their periodicity and their intrinsic luminosity. The more luminous the Cepheid variable, the longer it took to cycle between its maximum and minimum brightness.

This was critically important because it meant that you could calculate precisely how far away any Cepheid variable was by comparing its intrinsic luminosity against how bright it seems from our vantage point on Earth. The key question in astronomy at the start of the twentieth century was the distance to what were then called spiral nebulae, objects we now call galaxies. Some astronomers thought they were within the Milky Way Galaxy, others thought they were far outside it. Hitherto the only way to accurately measure the distances of astronomical objects was to use parallax, but in common with all but the closest stars, the spiral nebulae showed no parallax at all.

Leavitt discovered Cepheid variables in the Magellanic Clouds. From their peri-odicity Harlow Shapley was able to deduce their luminosity, and by comparing these against how bright they appeared on Earth, he could calculate their distance. The tricky bit was that the connection between distance and apparent brightness isn't a simple thing to determine, and early estimates of the distances of Cepheid variables were somewhat off. In the case of the Cepheid variables in the Small Magellanic Cloud, Shapley came up with a distance of around 94,000 light years, a little under half the modern estimate. But though inaccurate by today's standards, the number revolutionized astronomy. Spiral nebulae were clearly very distant objects far outside the boundaries of the Milky Way.

SAO 37734 (Gamma Andromedae, Almach)

Almach is a famous double star consisting of a bright orange star (magnitude 2.3) and a much less brilliant blue–green star (magnitude 4.8). Because of the differences in brightness, and the relatively narrow gap between them – about 10 arcseconds – Almach is considered a tough double to split using small aperture telescopes. Steady skies are crucial, but it is also important that the telescope has cooled down ade-quately, otherwise convection currents within the optical tube will cause the image to jump about unhelpfully. The fainter of the two stars is itself a double star system, but its two components, a pair of B-type main sequence stars, are so close together that they are not easily resolved (though certainly possible with very large aperture telescopes).

SAO 38592 (Beta Persei, Algol)

Algol, or Beta Persei, is probably the most famous variable star in the sky. There is a bit of debate over when it was recognized as a variable star though. Unlike Mira, SAO 129825, which varies dramatically in brightness, Algol doesn't vary that much, from magnitude 2.1 at its brightest to magnitude 3.4 at its dimmest across a period of 2.9 days. Ancient astronomers seem to have thought that this star was odd, and its Arabic name, *al-ġūl*, meaning 'the ghoul,' hints at a certain malignity. But ancient astronomers didn't explicitly mention changes in brightness, so although some argue this was the first variable star known to science, most astronomers give that credit to Mira.

Algol is an eclipsing binary star. It consists of a relatively cool K-type primary star (Algol A) and a much hotter B-type secondary star (Algol B). As the two stars orbit each other, they eclipse one another as seen from our vantage point on Earth. When the blue K-type secondary eclipses the yellow B-type primary, what is called a secondary eclipse, the brightness of Algol dips slightly. In fact the dimming is so slight that it cannot be detected visually. This is because it is the secondary that pumps out the largest portion of the light we see, so the fact the primary is being occulted by the secondary isn't of any great importance. But when the secondary is occulted by the primary, what is called a primary eclipse, the brightness of Algol drops dramatically.

SAO 38787 (Alpha Persei and the Alpha Persei Cluster)

Alpha Persei, also known as Mirfak, is a yellowy-white supergiant 593 light years away from Earth. At magnitude 1.8 it is the brightest star in Perseus, a distinctive but relatively dim constellation. Alpha Persei is noteworthy because it is a member of a star cluster known as the Alpha Persei Cluster. Because this cluster is relatively nearby, it doesn't look like much to us, and its members are spread out across more than 3° of space. Inevitably such an object is best viewed through binoculars or a rich-field telescope rather than a go-to SCT or Maksutov–Cassegrain.

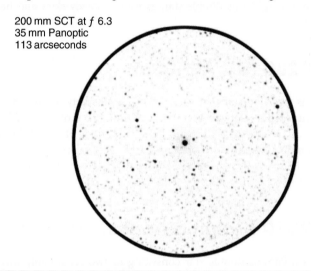

200 mm SCT at *f* 6.3
35 mm Panoptic
113 arcseconds

Figure 5.25. Alpha Persei is part of the Alpha Persei Cluster, a good object for observation through binoculars but rather less impressive through telescopes. (Image produced using Starry Night Pro. AllSky data courtesy of Main-Sequence Software Inc.)

SAO 39955 (Epsilon Aurigae)

Most of the time Epsilon Aurigae shines at magnitude 2.9, but every 27 years its brightness drops to magnitude 3.8, and remains at this minimum brightness for about a year. Between 2009 and 2011 the star will dim, remaining at its minimum for most of 2010. If you're using binoculars or a rich-field telescope, there are some stars in the neighborhood against which Epsilon Aurigae can be compared. Hoedus I (SAO 39966) and Hoedus II SAO 40026) shine at magnitudes 3.7 and 3.2, respectively. Hoedus I is therefore a trifle brighter than Epsilon Aurigae at its dimmest, while Hoedus II is a bit dimmer than Epsilon Aurigae at its normal brightness.

There are various theories to explain this dimming. One suggestion is that there is a star surrounded by a huge clump of dust that orbits Epsilon Aurigae, and as it passes between us and Epsilon Aurigae, the light from Epsilon Aurigae is considerably reduced.

SAO 58636 (Theta Aurigae)

Theta Aurigae is a tough double star for small to medium aperture telescopes. It consists of a blue–white, magnitude 2.6 star with a much dimmer, magnitude 7.1, yellow companion. An aperture of around 100 mm is generally regarded the minimum required to split this double star, as well as steady skies and high levels of magnification.

SAO 59280 (UU Aurigae)

UU Aurigae is a carbon star that varies in brightness between magnitude 7.8 and 10.0. Its brightness does not vary regularly, and though it is often described as having a periodicity of 232–234 days, the gaps between periods of peak brightness can be considerably more than this. Carbon stars like UU Aurigae contain a lot of carbon in their atmospheres, and this gives them an unusually red coloration when compared to many other stars.

SAO 92680 (Gamma Arietis, Mesarthim)

Mesarthim is a famous double star consisting of two remarkably similar fourth-magnitude blue–white stars. The two components are separated by 7.7 arcseconds and are easily resolved even with small aperture telescopes.

SAO 94027 (Aldebaran and the Hyades)

Aldebaran is one of the most famous stars, forming the bright orange eye of Taurus the Bull. It has numerous connections with mythology, astrology, magic, and folklore. The unmanned spacecraft *Pioneer 10* is traveling in the general direction of Aldebaran at more than 12 km/s; although that sounds terrifically fast, and by Earth-bound standards certainly is, at more than 26,800 miles/hour it will still take *Pioneer 10* about 2 million years to reach Aldebaran.

Aldebaran is an irregular variable star that varies in brightness between 0.75 and 0.95. It is also a multiple-star system with at least six components, though one of them (component E) has not been detected visually since 1899. The easiest component other than Aldebaran is component C, a magnitude 11.3 star almost 2.2 arcminutes away. This star isn't a companion of Aldebaran but simply in the same line of sight.

The remaining components are thirteenth magnitude stars are difficult to see past the glare of Aldebaran. One of these, Aldebaran-B, is a red dwarf companion star orbiting Aldebaran at a distance of around 650 astronomical units (i.e., 650 times the Earth-Sun distance). It is a little over 31 arcseconds from Aldebaran, but because of the enormous difference in brightness, detecting Aldebaran-B is incredibly difficult, even with a large aperture telescope.

From our perspective, Aldebaran appears to be a member of the scattered open cluster known as the Hyades. This is a line of sight illusion, though. Aldebaran is about 65 million light years away, whereas the Hyades are over 150 light years away. The Hyades themselves are an attractive group of a few hundred stars measuring 5.5° in angular diameter. Being so large, this cluster isn't usually observed using a telescope and is best appreciated with the naked eye or binoculars.

SAO 107436 (AG Pegasi)

AG Pegasi is a slow nova, an object that is, as its name suggests, a nova that takes much longer to brighten and fade than do regular novas. Until the mid-1850s AG Pegasi was a ninth-magnitude star, but across the next 20 years it slowly brightened up to around sixth magnitude. Astronomers noticed that it thereafter behaved like a variable star but over time becoming steadily dimmer. It now erratically varies between magnitudes 7.5 and 9.5.

SAO 110707 (Gamma Ceti)

Gamma Ceti is a challenging double star that cannot be resolved with small aperture telescopes. Even medium aperture telescopes will find Gamma Ceti a tough split, and it is primarily a double star for owners of telescopes with apertures upwards of 200 mm. Although they are not terribly close together, about 2.8 arcseconds, the primary star is much brighter (at magnitude 3.5) than the secondary (at magnitude 7.3), and it is this that makes Gamma Ceti a difficult split. On the plus side, the contrast between the yellow primary and blue secondary is very attractive, so like a lot of the tougher double stars, it's well worth the effort.

SAO 111674 (47 Tauri)

47 Tauri is difficult double star for medium aperture telescopes. The yellow magnitude 4.9 primary and its white magnitude 7.4 companion are 1.1 arcseconds apart.

There is also a third twelfth-magnitude component about half an arcminute away visible through large aperture telescopes.

SAO 125159 (Eta Aquilae)

Eta Aquilae is a variable star noted for being among the brightest Cepheid-type variable stars in the northern sky. Its brightness varies from 3.6 down to 4.4, with a period of just under 7.2 days. It is a fun addition to an observing program because of the proximity of another star, Iota Aquilae, SAO 143597, in the same part of Aquila. This star shines at magnitude 4.4, and by slewing between the two stars, it should be obvious whether Eta Aquilae is at its maximum or minimum brightness. The two stars are about 4.5° apart, so it's not possible to put them in same field of view except when binoculars or rich-field telescopes are used.

SAO 127029 (Epsilon Pegasi, Enif)

Epsilon Pegasi is an attractive double star that can be split with even small aperture telescopes. The primary is a magnitude 2.4 orange supergiant, while the secondary is a magnitude 8.4 blue dwarf. The two stars are separated by more than 2 arcminutes.

SAO 128318 (O'Meara's Little Ladle)

This star is in the middle of an asterism of stars in Pisces that Stephen James O'Meara refers to as the Little Ladle. The asterism is about 15 arcminutes in width, and looks good at medium magnifications. Using a 200-mm SCT, a Plössl eyepiece between 12 and 20 mm should do the trick nicely.

SAO 128374 (19 Piscium)

This is a bright carbon star easily viewed with telescopes of all sizes. Like all carbon stars, its red color is caused by clouds of carbon soot in the upper atmosphere of the Sun. Interestingly, it isn't quite so blood-red as some carbon stars, for example La Superba (SAO 44317).

SAO 129825 (Omicron Ceti, Mira)

Mira is a variable star that was first systematically observed by the German astronomer David Fabricius in 1596. It wasn't until 1638 that Johannes Holwarda, a Dutch astronomer, recognized that Mira wasn't simply an oddity but a star that fluctuated in brightness in a predictable way.

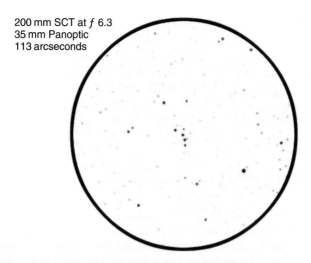

200 mm SCT at f 6.3
35 mm Panoptic
113 arcseconds

Figure 5.26. SAO 128318 is the brightest star in an asterism Stephen James O'Meara thinks looks like a ladle. (Image produced using Starry Night Pro. AllSky data courtesy of Main-Sequence Software Inc.)

Until the discovery of what we now call variable stars, astronomers assumed that stars shone with constant brightness. Mira clearly differed, its brightness varying from magnitude 3.4 down to magnitude 9.3 across an approximately 332 day cycle. This is a roughly 1,000-fold difference, taking it from being easily visible with the naked eye at its brightest to being too dim to see with the naked eye at its faintest. Discovering a star that varied as dramatically as this played an important part in the way astronomy (and science generally) was changing during the sixteenth and seventeenth centuries.

In fact the brightness of Mira doesn't vary in a completely regular way. On occasion it can be much brighter or much dimmer than expected. William Herschel, for example, recorded its peak brightness in November 1779 as being similar to that of Aldebaran, a star that shines at a magnitude of about +0.85.

Mira is actually a double star system. The primary star is called Mira A, and is a red giant with a diameter approximately 700 times that of the Sun. The secondary is called Mira B, and while it was long assumed to be some sort of white dwarf star, it is now recognized as a K-type main sequence star a little smaller than the Sun.

Until recently the closeness of Mira B to Mira A – about 0.6 arcseconds – meant that it couldn't be seen even in photographs. However, astronomers finally got to see Mira B when the Hubble Space Telescope took a fully resolved picture of the two stars in 1995. Given that Mira is only 400 light years away, the 0.6 arcseconds separate between the two components actually represents a remarkably small distance, a mere 70 astronomical units. To put this in perspective, the orbit of Pluto takes it from between 30 and 50 astronomical units from the Sun.

SAO 142996 (15 Aquilae)

15 Aquilae is an attractive double star consisting of a magnitude 5.4 yellow primary and a magnitude 7.0 secondary of debatable color. On paper at least this star should be yellow as well, though perhaps a slightly different shade. But there are numerous observers who report that the secondary star is purple or lilac in color (though admittedly the author is not among them). Regardless, this star is easily split even with small telescopes.

SAO 145457 (Beta Aquarii)

Beta Aquarii is also known by its Arabic name, Sadalsuud, which can be loosely translated as 'the luckiest of lucky stars.' Whether or not this star actually brings you good luck is not for the author to predict, but Beta Aquarii is an interesting star worth observing. It is a member of a rare class of stars called yellow supergiants, a giant version of the otherwise dwarf-sized F or G-type main sequence stars. The Sun, for example, is a G-type yellow dwarf, and Sadalsuud is broadly similar to the Sun in terms of composition and surface temperature, but with something like six times its mass.

SAO 146107 (Zeta Aquarii)

Zeta Aquarii is a double star consisting of two white stars of similar brightness (magnitudes 4.3 and 4.5 for the primary and secondary respectively). They are relatively close together, at about 2.5 arcseconds apart, but because the two stars have similar brightness, they are not difficult to split except perhaps with the smallest telescopes.

SAO 191524 (Fomalhaut)

Also known as Alpha Piscis Austrini, the star the Arabs named Fomalhaut, the Mouth of the Whale, is a notably bright (magnitude 1.16) star in a part of the sky otherwise devoid of bright stars. It is the eighteenth brightest star in the sky and only 25 light years away.

As part of the constellation of Piscis Austrinus, the Southern Fish, Fomalhaut never rises very far above the southern horizon for observers at far northern latitudes; at the latitude of southern England, for example, it only reaches an altitude of about 9.5°. Nonetheless, it is so bright that it shines through all but the worst light pollution and should be easy to see from locations with unobstructed horizons.

Fomalhaut is a young star, possibly a mere 200 million years old. It is of the same general type (and age) as Vega, and both stars emit unusually high quantities of infrared radiation. This appears to come from a disc of debris that encircles the star, probably a protoplanetary disc within which planets are forming. Indeed, in 2008 a planet was discovered orbiting the star just inside its protoplanetary disc. As the planet, Fomalhaut b, orbits Fomalhaut, it clears up any debris it encounters, and this is likely one reason for the gap between the protoplanetary disc and the star itself.

Fomalhaut will be a familiar name to those astronomers with a fondness for the weird fiction of H. P. Lovecraft and those who have written stories based on his Cthulhu mythos. Lovecraft mentions the star in several stories and poems, and August Derleth went on to cite Fomalhaut as the home to one of his better known contributions to the mythos, the fiery being Cthugha. Only when Fomalhaut is positioned properly in the night sky can Cthugha be summoned by his cultists... Will the stars be right next time you're outside observing this fascinating star?

Appendix: What You're Looking At

Early astronomers divided astronomical objects into three main types: planets, stars, and what were called mists or clouds; in Latin, *nebulae*.

Stars had fixed positions relative to one another, but the planets moved about, and so the ancient Greeks called them wanderers, *planētēs*. The changing positions of the planets were hard to explain, and none of the Solar System models proposed by the ancient Greeks or Romans adequately explained things such as retrograde motion. But even though they didn't understand why the planets moved about, the Greeks and Romans both placed huge value on interpreting the movements of the planets. This is the art of astrology, a practice that has developed independently in many different cultures around the world.

That stars were in some way similar to the Sun was something the ancient Greeks had guessed at, but it wasn't until the nineteenth century and the invention of spectroscopy that scientists could demonstrate that the light produced by the Sun was the similar to that produced by the stars. Over time different types of stars were identified, and it soon became clear that the Sun was a relatively small and ordinary star by cosmic standards, even though it is vitally important to us.

Nebulae posed particular problems for astronomers prior to the invention of the telescope. To the naked eye all nebulae look alike, nothing more than faint, misty patches of light. Only with the telescope was the diversity of nebulae revealed, but while they could be seen to be morphologically different, astronomers of the early modern era didn't understand that a planetary nebula is a very different thing to something like the Great Nebula in Andromeda, what we'd today recognize as a galaxy. Indeed, some astronomers viewed nebulae more as nuisances than as objects worthy of investigation. The deep sky objects Charles Messier added to his famous catalog were only of interest to him because they might be mistaken for comets, objects of far greater importance at the time.

N. Monks, *Go-To Telescopes Under Suburban Skies*, Patrick Moore's Practical Astronomy Series, DOI 10.1007/978-1-4419-6851-7,
© Springer Science+Business Media, LLC 2010

Eventually the differences between nebulae were understood, and galaxies in particular caused a profound rethink about the size of the universe. Once it became clear that these 'spiral nebulae' were galaxies like our own Milky Way Galaxy, astronomers had to accept that the universe was a much bigger place than they had previously thought. Just as the Sun turned out to be a very ordinary star, so the Milky Way was revealed as just one galaxy among millions.

Exploring the night sky with a go-to telescope lets you condense thousands of years' worth of scientific progress into an hour or two. As the preceding chapters have hopefully revealed, even objects that appear quite unexciting visually become much more interesting once you understand why they're significant.

Stars

What Are They?

Stars are essentially very large, very hot balls of gas, primarily hydrogen and helium. Their enormous mass means that their own gravity pulls them into a spherical shape, and the immense pressure and temperature this creates within them initiates nuclear fusion. Hydrogen atoms are fused together to make helium, and as this happens energy is released in the form of heat and light.

Stars vary considerably in color and temperature, from around 50,000 K down to a mere 3,000 K. Hot stars emit more of their light towards the blue end of the spectrum; in the cases of very young and hot stars they may even be emitting a lot of ultraviolet light human observers cannot even see. Cooler stars emit more light towards the red end of the spectrum, and in fact many cool stars emit most of their light in the form of infrared.

So although all stars emit light across the color spectrum, they emit light most strongly at certain wavelengths, depending on their temperature, and that in turn is reflected in their color. Astronomers classify stars according to their color, from hottest to coolest as spectral types O, B, A, F, G, K, and M. Of these, O-, B-, and A-type stars appear blue to white, F- and G-type stars are white to yellow, and K- and M-type stars are orange to red. Astronomers further subdivide these spectral types into ten classes, for example G0, G1, G2, G3, and so on through to G9. So a G1 star would be cooler than a G0 star, but a G9 star would be hotter than a K0 star.

There are some unusual star classes as well. The most important of these from the perspective of backyard astronomers are carbon stars (sometimes called C-type stars) and Wolf-Rayet stars (sometimes called WR-type stars).

Carbon stars are deep red stars that are even cooler than the orange M-type stars. Partly they're red because they're cool, but they're redder than M-type stars because their atmospheres contain a lot of carbon. The heat and pressure inside a star fuses hydrogen into helium, but it can also synthesize other elements as well, including carbon and oxygen. Carbon stars are so old that they have accumulated a lot of carbon, and this comes to surface and forms patches of soot in the atmosphere. In the same way that dust in Earth's atmosphere makes the setting Sun look redder than it does high up in the sky, this dust in the atmosphere of a carbon star scatters blue light while letting red light pass through, giving a carbon star its distinct red appearance. The patches of soot aren't stable, and come and go on an irregular basis. In doing so the brightness of the carbon star varies, and carbon stars are variable stars, about which more will be said shortly.

Wolf-Rayet stars are supergiant stars approaching the end of their life. Having used up most of their hydrogen, helium and other elements are being fused together. This process produces enormously high temperatures, and Wolf-Rayet stars are among the hottest known stars. Gamma Velorum, for example, has a surface temperature of 50,000 K. However, the tremendous heat and radiation produced by Wolf-Rayet stars creates winds moving up to 9 million km/hour, and needless to say this blasts off huge amounts of material into space. Indeed, what we see when observing Wolf-Rayet stars isn't usually the star itself but the winds around it. Within a relatively short period of time, likely just a few million years, the fuel that powered this process is exhausted, and the star collapses into a planetary nebula or a supernova.

How Big Are They?

A star has to be above a certain size simply to have enough mass for nuclear reactions at its core to begin. Gravity comes from mass, and it's gravity that squeezes the core of a star and thereby creates the incredible temperatures required for nuclear fusion to take place. At the core of a star such as the Sun the temperature will be around 15,700,000 K!

The more massive a star, the hotter its core will be, and this means that nuclear reactions will proceed faster than in a less massive star. This brings us to a key thing about star sizes: the bigger the star, the shorter its life will be.

Stars can be broadly divided into those stars that are 'giants' and those that are 'dwarfs.' The Sun is a dwarf star, while giants can be hundreds of times its mass. Astronomers normally divide stars into seven size categories, using Roman numbers running from I for the biggest stars (supergiants) through to VII for the smallest (white dwarfs).

So, by combining the spectral type (i.e., the color of a star) with its size, you can quickly describe a star very accurately. This tends to be how professional astronomers describe stars. The Sun would be designated as a G2V star, i.e., a G-type star, hotter than a G3 star but cooler than a G1 star, and in terms of size a category V star, i.e., a dwarf.

Alpha Centauri A is another G2V star, and its similarity to the Sun is one reason why it is so interesting. Sirius is an A1V star, a star similar in mass to the Sun, but very much hotter. On the other hand, the famous Garnet Star, Mu Cephei, is designated as a M2I star, a star that is not only cooler than the Sun but also much more massive.

How Bright Are They?

Some stars appear brighter than others because they are nearby. Both the Sun with Alpha Centauri A are G2V stars of similar size and temperature, and they both give out about the same amount of light. But the Sun is a far brighter object from our vantage point on Earth simply because it is very much closer. Whereas Alpha Centauri A is 41,300,000,000,000 km away, the Sun is 276,000 times closer at a distance of just 149,598,000 km.

Put another way, stars of similar apparent brightness can be situated at very different distances. Vega and Rigel seem equally bright when viewed from Earth, but Vega is a much closer star, just a mere 25 light years away, whereas Rigel is about 800 light years away. But because Rigel is intrinsically a much more luminous object than Vega, the two stars appear to be similar in brightness from our point of view.

Astronomers therefore need to describe the brightness of stars in two different ways. The first is to describe their apparent magnitude, which is how bright they seem to us. The second measurement is their absolute magnitude, which is how bright a given star would seem when observed from a set distance (in this case 10 parsecs, or about 32.6 light years).

The apparent magnitude system goes back to the times of the ancient Greeks, who would classify the brightest stars as being among the first magnitude, the next brightest as the second magnitude, and so on.

The system was standardized by Norman Robert Pogson in 1856 by placing Vega at the zero point. Stars brighter than Vega have negative values, as with Sirius (−1.46) while stars that are dimmer than Vega would have positive values, as with Deneb (1.25). The differences between the magnitudes was standardized as well, when it was determined that a typical sixth magnitude star had just one hundredth of the brightness of a typical first magnitude star. Although the system has been tweaked over the years, the apparent magnitude system remains the most widely used system among amateur astronomers (and is the one used throughout this book).

The absolute magnitude works in the same basic way, but it is far more useful to professional astronomers because it allows the intrinsic brightness of different stars to be compared more easily. The Sun, for example, has an absolute magnitude of 4.85. That's about the same as a very unremarkable night sky star such as 15 Orionis or 66 Ophiuchi (SAO 94359 and SAO 123005). On the other hand, Rigel has an absolute magnitude of −6.7, making it one of the most intrinsically luminous stars in the Milky Way galaxy, and more than half a million times brighter than the Sun.

How Old Are They?

The color of a star and its spectral class is really just a snapshot of its current state. An average star such as the Sun will be a yellow dwarf for most of its life, placing it within a group of stars astronomers refer to as main sequence stars. But for the last few percent of its life it expands in size dramatically, leaving the main sequence and becoming a red giant. Arcturus is the fourth-brightest star in the sky and the brightest red giant star in the sky.

Main sequence stars become red giants because the nuclear reactions in the core of the venerable star change, causing the outer layers of the star to expand. Red giants are very much larger than the stars they evolve from. A yellow dwarf star such as the Sun will expand into a red giant with a radius equal to the distance of Earth to the Sun. As well as becoming much larger, red giant stars are also much cooler than the dwarf stars they evolved from. Where the Sun has a surface temperature of approximately 5,800 K, the red giant it is likely to have a surface temperature of only 3,500 K or so.

The time it takes for a main sequence star to become a red giant star depends upon its mass. The bigger the star, the faster it consumes its reserves of nuclear fuel, and the more quickly it will become a red giant. A small star like the Sun will take about 10 billion years to work through its main sequence phase. But a more massive star will have a much shorter lifespan; Beta Andromedae, for example, has a mass of about 5 times that of the Sun, and will be a main sequence star for less than 200 million years.

Very massive stars do not become red giants; they become red supergiants instead. Betelgeuse is the classic red supergiant, a star with about 20 times the mass of the Sun that has inflated into a gigantic object around 1,000 times the diameter of the Sun. Red supergiants are even cooler than red giants, with a surface temperature of around 3,000 K.

Blue supergiant stars differ from red supergiants in being hot rather than cool. But like red supergiants, blue supergiants are dying stars. Although tens of times more massive than the Sun, red supergiant stars are millions of times larger in terms of volume, and that means their density is very low. The gravitational pull of the core isn't enough to keep the outer layers from drifting away into space, and if a red supergiant star loses enough material the nuclear reactions inside slow down and the light they emit shifts from being strongest at the red end of the spectrum to being strongest at the blue end. Rigel is the most famous blue supergiant.

Neither red nor blue supergiants are stable in the long term. Sometimes stars shuffle between the two states. Polaris is an example of a supergiant star evolving from a blue supergiant back into a red supergiant, and consequently it has a yellow color intermediate between the two. But eventually the nuclear fusion in the center of a supergiant star will stop producing enough energy to maintain its bloated size. When that happens the star collapses in on itself, becoming a supernova.

How Far Away Are They?

Stars are, of course, a long way away, at least by the standards of present-day space travel. Proxima Centauri is the closest star to us, a mere 4.22 light years away. The fastest manned spaceship to date is *Apollo X*, which at one point reached a speed of just over 11 km/s. That works out at about 350,000,000 km/year, but even at that speed it would take *Apollo X* nearly 114,000 years to reach Proxima Centauri.

Because these numbers are so big, a very long unit of length is used instead, the light year. A light year is simply the distance that light covers in a year. A light year is about 10,000,000,000,000 km. This is about 63,000 times the distance from Earth to the Sun, a unit conveniently referred to as one astronomical unit. Light covers one astronomical unit in about 8.33 minutes, but it takes light 4.22 years to cover the distance between us and Proxima Centauri.

Professional astronomers often use another unit instead, the parsec. This is the distance an object must be to have a parallax of 1°. The parsec was invented by Herbert Hall Turner in 1913 and is useful because the distances between Earth and the stars were then (and to some degree still are) determined by using parallax. This is the optical illusion that causes the position of a nearby star to shift very slightly relative to stars in the background as Earth orbits the Sun. It's a comparable illusion to how a finger held at arm's length seems to move if you alternately close one eye and then the other.

As it turns out, a star would need to be about 3.26 light years to have a parallax of 1°. There aren't any stars at precisely this distance, though, and even the closest stars exhibit parallaxes of less than 1°. Indeed, most stars in our galaxy are so far away that their parallax is so small it cannot be measured accurately. But still, parallax has been useful in determining the distances of the closer stars, and these in turn provided baseline values against which newer, more wide ranging techniques could be used to calibrate distances.

The most notable of the later techniques for determining distances is that involving Cepheid variables. These stars are often called 'standard candles' by writers describing the history of astronomy. The circumpolar star Delta Cephei was the first described, and what makes these objects special is that there is a tight relationship between their intrinsic luminosity and the rate at which their brightness varies. Since there's also a relationship between how bright a star appears and how far away it is, if you compare its apparent brightness to its intrinsic luminosity, you can estimate its distance. It's actually a bit more complicated than that, but the upshot was that Cepheid variables provided astronomers with a reasonably reliable way to measure the distances to things such as galaxies that couldn't be deduced using parallax.

What Can You See Through a Telescope?

Some stars are worth observing simply because they are unusually colorful; these include stars such as the very blue star Bellatrix and the very red star Mu Cephei. Others, such as the very Sun-like Eta Cassiopeiae, are objects that might not be particularly dramatic but are interesting nevertheless. But the stars that are most popular among amateur astronomers are those stars known as variable stars and double stars.

Variable stars exhibit variations in their brightness. Two famous examples are Mira and Algol, but there are lots of others. By comparing the brightness of a variable star against some fixed brightness stars in the immediate area, the brightness of the variable star can be estimated. Many amateur astronomers routinely survey dozens of variable stars during their observing sessions, and the data collected in this way can be used to plot the brightness of a variable star against time, a plot known as a light curve.

Observing variable stars is challenging and requires a good eye, since determining the brightness of a star by comparing it with others is a skill that develops over time. Because of this, variable stars are an aspect of the hobby many casual astronomers tend to overlook, but ironically it's also one part of the discipline of astronomy where amateurs are just as important as professionals. By collecting data on numerous variable stars over long periods of time, dedicated amateurs generate the raw data the scientists need to work with. In doing so, amateurs are not just enjoying astronomy but helping to push its boundaries outwards. The American Association of Variable Star Observers is probably the best-known organizer of this type of work.

Double stars contain two or more stars in close proximity. There are two sorts, optical doubles and true binary systems. Optical doubles are simply line of sight illusions, and the stars involved may be many light years apart. Binary systems involve two or more stars that are gravitationally bound to one another. In the case of two-star systems, the brighter star is usually called the primary and the dimmer star the secondary. Where more than two stars are involved, each component will be given a letter; typically, but not always, an A is given to the brightest, B to the next brightest, and so on. A good example is 15 Monocerotis, a pretty triple star in Monoceras. Component A is a magnitude 4.7 star, component B a magnitude 5.2 star, and component C a magnitude 6.1 star. Sometimes component stars within binary systems are discussed by themselves, in which case the letter is added to the name of the system, so in this case you might refer to the brightest star in the trio as 15 Monocerotis A.

Whereas observing variable stars can be put to good practical use, double stars are generally observed simply for fun. There are two ways to enjoy them. The first is simply because they're pretty. The color contrasts between two adjacent stars is often pleasing, but sometimes it's downright unexpected. Gamma Andromedae and Epsilon Boötes are two famous examples of double stars that contain a blue star that appears green, at least to some people, likely because of the specific color contrast between the two components.

The other fun way to enjoy double stars is an objective test of your eyesight or optical equipment. Many double stars are 'tight,' meaning the two components are so close together they aren't immediately resolvable. At high magnifications the two stars may be split, but this will depend on the aperture of the telescope. Small aperture telescopes have less resolving power than large aperture telescopes, and stars that can't be split with a 75 mm telescope may well be split without problems using a 200 mm telescope.

What's the Best Way to View Them?

Stars are good targets for both small and large aperture telescopes. Small aperture telescopes often show their colors rather better than large aperture telescopes and are more than adequate for observing variable stars and splitting many excellent double stars. Large aperture telescopes give you more resolving power, so tighter doubles can be split, but star colors can sometimes seem a bit washed out.

When splitting double stars, reasonably high levels of magnification are often required, upwards of ×100. But because these objects have a small angular diameter, generally well under an arcminute, a wide field of view isn't required. In theory at least, a standard Plössl eyepiece is just as good for splitting tight double stars as wide-field eyepiece of similar focal length.

Having said that, short focal length Plössl eyepieces have very little eye relief, which means your eye has to be very close to the lens to see the image, and this makes it very uncomfortable to use. Using a Plössl eyepiece with a focal length shorter than 10 or 12 mm can be a real chore, so consider using a Barlow lens to halve the focal length of a more comfortable eyepiece such as a 20-mm Plössl. Some eyepiece manufacturers produce short focal length eyepieces that are specifically engineered to have lots more eye relief than standard Plössl eyepieces of similar focal length. The Televue Radian series and Vixen Lanthanum are two of the best known eyepieces of this type.

Planetary Nebulae

What Are They?

Planetary nebulae are left behind after a red giant (rather than a supergiant) star falls apart; they are therefore the fate of stars such as the Sun. The process that causes a star to expand into a red giant eventually allows the outer layers of the star to drift off into space. This material eventually cools down, but before it does so it emits light, and the glowing material around the remains of the old star are what we see as a planetary nebula. All that's left of the star itself is a small, faint object called a white dwarf.

Although they often seem small when viewed through a telescope, planetary nebulae are comparatively large objects; the Ring Nebula, for example, is well over a light year in diameter, even though the bright part we can see easily is barely an arcminute across. On the other hand, a red giant only loses about 10% of its mass to the nebula, so the nebula is actually exceedingly tenuous in terms of density. Indeed, a scoop of planetary nebula material wouldn't be very different to a laboratory-grade vacuum. Furthermore, it doesn't take long for a planetary nebula to dissipate completely, and most probably only survive for a few thousand years. In astronomical terms then, these objects are ephemeral.

The white dwarf star at the center of a planetary nebula is interesting. While a white dwarf retains around 90% of the mass of the red giant, it is much, much smaller. A star that started off with a mass similar to that of the Sun will swell up into a red giant and then collapse into a white dwarf about equal in size to Earth. In other words, a white dwarf packs 90% of the mass of the Sun into a space less than a millionth of its size.

Since they no longer perform nuclear fusion, white dwarfs don't generate heat and light any more, but it still takes them a long time to cool down to the background temperature of interstellar space. Indeed, because it takes so long for them to cool down, the universe hasn't been around long enough for any of them to cool down completely.

How Big Are They?

Planetary nebulae are relatively large objects compared to the stars that form them, but by the standards of deep sky objects such as star clusters they are rather small. A typical example like the Ring Nebula is about a light year across. Since they are formed from spherical shells of gas flying away from dying stars, planetary nebulae expand over time, but the bigger the nebula becomes, the fainter it appears. This is partly because the glowing gas is becoming more thinly spread out, and partly because the gas is cooling down and so emits less light. In fact what amateur astronomers usually observe is only the innermost portion of the nebula, where the gas is hottest and densest. The Ring Nebula, for example, would appear to be more than twice its size if the fainter, more attenuated shells of cooler gas could be seen as easily as the hotter core.

How Bright Are They?

Planetary nebulae are generally faint objects, but surprisingly perhaps, the ease with which they can be observed is not necessarily proportional to their apparent magnitude. The Ring Nebula and the Helix Nebula demonstrate this phenomenon nicely. The Ring Nebula has an apparent magnitude of 9.5, while the Helix Nebula is a magnitude 7.3 object. These numbers would seem to suggest that the Helix Nebula is a much brighter object, and therefore much easier to see, but this is not the case. The Ring Nebula is an easy target, even for small telescopes, while the Helix Nebula is notoriously difficult to see even with a big telescope.

In fact the ease with which a planetary nebula can be seen – and indeed deep sky objects generally – comes down to a combination of both apparent magnitude and angular diameter. A faint but small object will often be easier to pick out from the background sky – in other words, show more contrast than a brighter but much larger object. Astronomers often describe this phenomenon as overall surface brightness, and the Helix Nebula at more than 25 arcminutes in diameter is the classic example of a large object with a very low surface brightness. The Ring Nebula, on the other hand, is barely an arcminute across, so what little light it does emit is so much more concentrated that the object is very much easier to see.

How Old Are They?

The hot gas that makes up a planetary nebula quickly cools down and emits visible light for only a few thousand years.

How Far Away Are They?

The closest planetary nebula is the Helix Nebula, a mere 700 light years away, and this is why it seems so large when viewed from Earth compared with the much more distant Ring Nebula 2,300 light years away. Planetary nebulae more than a few thousand light years away are generally too small and faint to be seen through the usual telescopes used by amateur astronomers.

What Can You See Through a Telescope?

Amateur astronomers usually view the glowing gas part of the nebula, but in some cases large aperture telescopes will reveal the white dwarf at the center of the nebula as well.

The shape of a planetary nebula largely depends upon its orientation relative to our viewpoint. Although most planetary nebulae are essentially spherical in shape, the distribution of gas within them isn't uniform, and instead the densest regions tend to form two cone-shaped structures, one above the north pole of the star and the other below the south pole. Viewed from above one of the two poles, a planetary nebula usually looks circular, since we're looking downwards into one or other of the cones. This is the case with the Helix Nebula and the Ring Nebula, both of which have a more or less circular shape. But when a planetary nebula is viewed from the side, the two cones can look somewhat like a bow-tie, as is the case with the Dumbbell Nebula.

Planetary nebulae are unusual among deep sky objects in that they are sometimes quite colorful. Through large aperture telescopes especially, the brighter ones often appear to be either blue or green.

In most cases the white dwarf star is very faint, typically magnitude 15 or fainter. Only large aperture telescopes will reveal such stars. But a few planetary nebulae contain much brighter stars at their centers, the most notable perhaps being the magnitude 10 star at the heart of the Southern Ring Nebula (NGC 3132) in Vela. Another planetary nebula around a relatively bright (magnitude 11.4) star is NGC 1360 in Fornax. Unfortunately for far northern observers, these are both southern sky objects.

What's the Best Way to View Them?

As is usually the case with deep sky objects, the more aperture you can use when viewing planetary nebulae, the better. The Helix Nebula is something of an exception here, and can look quite good when viewed through relatively small aperture rich-field refractors and short focal length reflectors, but only under pitch-black skies. Under even slightly light-polluted conditions this object is notoriously difficult to see, almost regardless of aperture.

In most cases, though, planetary nebulae are objects best viewed with medium to large aperture telescopes at medium to high magnification levels. The aperture brings in the light, while increasing the magnification improves the contrast between the nebula and the background sky. Depending on the nebula, anything from ×100 to over ×250 may be in order. It's certainly worth experimenting between different eyepieces to find out which particular one works best.

Light pollution filters really pay for themselves when it comes to observing planetary nebulae. General purpose broadband filters are of some use, but narrowband and O-III filters are the two filters most astronomers find to be of the most use, the narrowband filter perhaps being the slightly more versatile and therefore better value. Since planetary nebulae are viewed at moderate to high magnifications, be sure to buy a filter that will fit your medium to short focal length eyepieces.

Supernova Remnants

What Are They?

A supernova remnant is simply what is left behind by a supernova. The classic example is the Crab Nebula in the constellation of Taurus.

There are two contrasting ways supernovae occur. The first involves a very large star running out of nuclear fuel such that nuclear reactions stop, and the second involves a white dwarf star accreting enough material from a companion star such that nuclear reactions begin.

With the first type, the star collapses because the pressure at the core is no longer sufficient to maintain the star's shape, and as it collapses into a neutron star or black hole it radiates massive amounts of energy. The second type of supernovae work differently, because it is a series of runaway nuclear reactions that cause the star to explode, but the effect is much the same.

In either case the resulting supernova is exceedingly bright, and for a short period of time, typically a few weeks, a supernova can be brighter than all the other stars in the galaxy put together. Supernovae are exceedingly rare, though; per century, only one or two will occur in a galaxy the size of the Milky Way, and the last one visible to the naked eye was discovered in 1604.

How Big Are They?

Supernova remnants tend to be bigger than planetary nebulae. The Crab Nebula, for example, is about 11 light years across, and the Veil Nebula is even bigger, around 100 light years across.

How Bright Are They?

Although supernova are very bright indeed, supernova remnants are mostly dim and difficult to see.

The Crab Nebula is informative in this regard. The supernova that formed it was observed by Chinese and Arab astronomers in 1054, who kept detailed records of such things. Apparently this 'new star' was visible as a nighttime object for 653 nights, and for 23 days it was visible even when the Sun was above the horizon, suggesting it was even brighter than Venus. Modern astronomers estimate that at its peak it had an apparent magnitude of around –6, or about 3.6 times brighter than the maximum brightness of Venus.

How Old Are They?

Supernova remnants don't hang around for long. The explosive forces that make them so bright initially also mean that they're expanding very rapidly, and it doesn't take long for the debris to be spread out so thinly it isn't visible any more. Astronomical records suggest Chinese astronomers observed a supernova in A.D. 185, an object known today as RCW 86 in the constellation of Circinus. Given its distance from Earth, astronomers believe that this supernova exploded about 2,000 years ago. Among modern surveys the oldest supernova remnants are around a million years old.

How Far Away Are They?

Amateur astronomers mostly observe relatively nearby supernova remnants in our own galaxy. The Veil Nebula for example is one of the closest supernova remnants, only 2,500 light years away. The Crab Nebula is a bit farther away, at around 6,500 light years distant.

Supernovae themselves have been observed in other galaxies, such as SN 1885A in the Andromeda Galaxy, and in theory at least supernovae in the nearer galaxies might well be detectable through amateur telescopes. By their nature these are transient phenomena, so rather than books you'll need to rely on astronomy magazines and websites to alert you to any such objects.

What Can You See Through a Telescope?

Supernova remnants aren't terribly impressive objects. They mostly look like faint, misty patches of light with tendrils around their edges. Relatively young remnants, like the Crab, are more or less elliptical in shape, but as they expand, the central part becomes less and less distinct, until eventually they acquire a more ring-like shape, like that of the Veil Nebula. In fact only segments of the Veil Nebula can be seen easily, the two brightest portions known as the Eastern Veil and (to a lesser degree) the Western Veil.

What's the Best Way to View Them?

Because they vary in size, a magnification that works well for viewing one remnant might not work so well on another. Taking the Crab and the Veil as examples, the Crab is quite a small object a mere 6 arcseconds across, while the Eastern Veil alone spans a good 2 arcminutes! So a certain amount of trial and error will be involved.

Light pollution filters are very useful when observing supernova remnants. The Veil Nebula is notorious for being one of those objects difficult to see without a filter, but quite straightforward when an O-III filter is used. The Crab Nebula is generally not too difficult to see under suburban conditions, but a narrowband filter does make it easier to pick out the details.

Reflection Nebulae

What Are They?

Reflection nebulae are clouds of molecular hydrogen and dust that reflect starlight. More specifically, what astronomers call 'dust' isn't much like the dust we find on Earth. Instead it is mainly spindle-shaped grains of ice and various minerals similar in size to the particles that make up cigarette smoke, about 0.001 mm across. When light hits these particles the blue light in particular is scattered, so that in photographs reflection nebulae typically appear blue. Unlike an emission nebula, the amount of energy they receive from the nearby star isn't enough to cause the gas and dust to ionize, i.e., glow.

Relatively few objects observed by amateur astronomers are purely reflection nebulae, though. Probably the best of the straight reflection nebulae are those associated with star clusters, since these are relatively easy to pull into the field of view, even if the nebula itself isn't easy to detect. The pick of the bunch in terms of the ease with which it can be seen is probably the Iris Nebula (NGC 7023) in Cepheus. The reflection nebula associated with Pleiades star cluster is perhaps better known, but much more difficult to see, and essentially invisible from suburban locations.

In many cases a reflection nebula is associated with an emission nebulae. The classic example is the Trifid Nebula (NGC 6514) in Sagittarius. The emission nebula is in the center and generally very easy to see, while the outer reflection nebula is notoriously difficult to spot.

How Big Are They?

Most reflection nebulae are fairly big, a few light years in diameter. The Iris Nebula, for example, is about 4 light years in width. In fact because these objects are composed of material drifting through the interstellar medium, their apparent size is more about the luminosity of the star (or stars) associated with the nebula. The more luminous the star, the larger the patch of gas and dust it will illuminate.

In terms of angular diameter these objects vary, but most are measured in arcminutes rather than arcseconds. They are therefore objects to be observed with medium rather than high magnification eyepieces.

How Bright Are They?

Reflection nebulae are notoriously difficult to observe except under pitch-black skies. Their large size means that they tend to have a very low surface brightness, so there isn't much contrast between the nebula and the background sky. Furthermore, because they reflect starlight, narrowband and line filters aren't of much help, since these block starlight. Broadband filters can be helpful, but only if the ambient light pollution isn't severe. In other words, even with light pollution filters in your toolbox, these aren't easy objects to observe from suburban locations.

How Old Are They?

Because they are simply clouds of gas and dust that happen to be close to a star, reflection nebulae don't really have ages or life spans in the same way as, for example, a planetary nebula. The reflection nebulae associated with the Pleiades will only glow for as long as it passes by these stars.

How Far Away Are They?

Being faint objects at the best of times, most of the reflection nebulae observed by amateur astronomers are relatively nearby. The Iris Nebula is typical, at a distance of 1,400 light years from Earth.

What Can You See Through a Telescope?

Simply detecting a reflection nebula can be challenging, particularly under suburban sky conditions. Under perfectly dark conditions averted vision may reveal some detail, though, typically a brighter core and faint tendrils around the edges.

What's the Best Way to View Them?

Suburban astronomers will inevitably find these objects disappointing because the best way to view them is to do so under dark, rural skies! This is why relatively few are mentioned in this book and none qualify as showpiece objects. Exurban astronomers may find them a bit more rewarding, particularly if a large aperture telescope is used. A broadband light pollution filter can help.

Emission Nebulae

What Are They?

Like reflection nebulae, emission nebulae are clouds of dust and gas. But unlike reflection nebulae, the amount of energy emission nebulae received from nearby stars is sufficient to cause the gas (mostly hydrogen) to become ionized, and so the nebula doesn't just reflect starlight; it emits some light of its own. This makes emission nebulae much brighter objects than reflection nebulae, and this is why they are so much easier to observe.

Emission nebulae include some of the most notable objects in the deep sky, including quite possibly the best of them all, the famous Orion Nebula.

How Big Are They?

Emission nebulae can be quite large. The Orion Nebula, for example, has an angular diameter of a little over a degree, which corresponds to a linear width of about 25 light years.

How Bright Are They?

Although some emission nebulae are quite bright, the Orion Nebula and the Trifid Nebula being two classic examples, the majority are comparatively faint objects. The problem is that their large angular diameter means that the light they emit is spread out across a wide patch of sky, so their overall surface brightness tends to be low. On the other hand, because they contain gas emitting light at specific wavelengths, these are ideal objects for observation using light pollution filters. Broadband filters, narrowband filters, and line filters all have their place here.

How Old Are They?

As with reflection nebulae, the clouds of gas and dust that make up an emission nebulae can be very ancient. However, emission nebulae are sometimes called stellar nurseries because they contain the right density of gas and dust for star formation to occur. But once these new stars spring into life, the light they emit starts driving away the material that made up the nebula in the first place. Over time the nebula dissipates, and nothing remains but a collection of hot, young stars – an open star cluster.

So although the clouds of gas and dust that become emission nebulae can be very ancient, the emission nebulae themselves tend to be quite young. Since they only emit light when excited by the radiation given off by the young stars within them, an emission nebulae won't be any older than the stars it contains. The Orion Nebula, for example, seems to be about a million years old.

How Far Away Are They?

The Orion Nebula is the closest emission nebula to Earth, at a distance of about 1,300 light years. The Trifid Nebula and the Lagoon Nebula are both a bit farther out, at about 5,000 light years away.

What Can You See Through a Telescope?

At their best, emission nebulae are wonderful objects to look at, with plenty of detail available. The showpiece emission nebulae such as the Orion Nebula will reveal not only their general shape but also details such as embedded star clusters and even hints of blue or green coloration. It's well worth switching between different focal length eyepieces to see what becomes apparent at different magnifications.

What's the Best Way to View Them?

Light pollution nebulae are particularly useful tools when observing emission nebulae. Broadband filters boost the contrast a bit, making it easier to detect the fainter edges of an emission nebulae. In the case of the Orion Nebula, for example, a broadband filter makes the full extent of the nebula easier to see, but has the advantage of not cutting out the starlight from embedded stars such as the famous Trapezium Cluster.

Narrowband filters will dim starlight, but they improve the contrast even further. Under suburban skies especially, a narrowband filter is extremely useful, even when quite bright nebulae are being observed. Among the line filters, the O-III filter is often a useful adjunct to a narrowband filter. Although a narrowband filter usually does the best job in terms of showing the full extent of an emission nebula, the two filters often improve the contrast of different portions of the nebula. Swapping between the two of them helps you to tease out different details, resulting in a better appreciation of the nebula's overall shape and structure. In the case of the Orion Nebula, for example, an O-III doesn't do much to improve the view of the tendrils around the edges of the nebula, but the structure of the inner regions of the nebula tend to be a bit easier to see.

H-beta filters aren't as generally useful as either narrowband or O-III filters, but in some cases they reveal details a little better than the other filters. The M43 part of the Orion Nebula, for example, is a noteworthy example of an emission nebula that reveals additional detail when observed with an H-beta filter.

Because they have large angular diameters, it's tempting to use very low power eyepieces when observing emission nebulae. Up to a point, that's fine, but exit pupil can be a critical issue here. The exit pupil of a given telescope/eyepiece combination has to be less than the size of the pupils of the observer. If the exit pupil is wider than the pupils of the observer, not all the light will get into the eye, and the user ends up wasting some of the light collected by the telescope. Most children and young adults have pupils that expand to about 7 mm under very dark conditions, while people above the age of 30 tend to have pupils that expand rather less, down to about 5 mm in the case of seniors.

The exit pupil is the aperture of the telescope divided by the magnification delivered by the eyepiece currently being used. So a 200 mm telescope at ×40 (as with a 50 mm Plössl) will have an exit pupil of 5 mm. That's fine for a fully dark-adapted eye. But throw a reducer-corrector into the mix, and the 50-mm Plössl will yield a magnification of ×25, and the exit pupil will be increased to 7.9 mm. That's larger than a fully dark-adapted eye, and consequently this particular hardware set-up wouldn't be ideal.

Open Clusters

What Are They?

Open star clusters are loose associations of stars bound together by mutual gravitational attraction. They are made up of stars born from the same nebula. Although usually traveling in different directions, the combined gravitational pull of the stars still binds them together somewhat. This pull isn't strong enough to keep them bound together permanently, and eventually the stars will drift away along their own paths. But for a period of time at least, they are sufficiently slowed down that they form a discrete cluster.

Because open star clusters are mostly found along the plane of the Milky Way, astronomers sometimes call them galactic clusters. This contrasts them with globular clusters, which are mostly in a halo surrounding the galactic core. The two constellations where open star clusters are found in greatest abundance both straddle the Milky Way, Cassiopeia in the north and Puppis in the south. In either case observers are looking along the galactic plane but away from the galactic core. Unlike galaxies, which of course lie far outside the Milky Way Galaxy, few open clusters are found substantially above or below the galactic equator.

How Big Are They?

Size varies considerably, but most are a few tens of light years across. The two open clusters that make up the Perseus Double Cluster are towards the high end of the size range, a little over 60 light years across each.

Open star clusters can contain anything from a few dozen stars through to thousands of them. Most of the ones observed by amateur astronomers contain around 100–150 stars, at least when viewed through medium to large aperture telescopes under dark skies. There are some exceptions, though. At the low end of the range is M29 in Cygnus, a cluster that seems to contain about 50 members when viewed from our location in space. In fact it probably contains many more stars than this, but a passing cloud of dust in between us and the cluster blocks out the light from many of its members. Although M29 is notoriously unimpressive, M11 in Scutum is often considered to be one of the most impressive open star clusters. It contains over 1,000 stars and is so densely packed that it can easily be mistaken for a globular cluster.

How Bright Are They?

Open star clusters run the range from easy naked-eye objects such as the Pleiades through to challenging objects such as NGC 2158 in Gemini that will only be resolved with large aperture telescopes. On the whole, though, they are good objects for small to medium aperture telescopes and bright enough to be relatively forgiving in terms of light pollution.

How Old Are They?

By their very nature, open star clusters tend to be young objects. As time passes the stars that make them up drift apart, and eventually the open star cluster simply ceases to be, its constituents being so far away from each other that there are minimal gravitational interactions between them. In fact most stars were probably members of a open star cluster at some time or another, including our own Sun.

Most open star clusters are just a few tens of millions of years old. The Pleiades, for example, is about 100 million years old. But some are substantially older than this; M11 in Scutum, for example, is about 200 million years old, while the Hyades are upwards of 600 million years old. Only very few approach the sorts of ages typical of globular clusters, though. NGC 188 and NGC 6939, both in Cepheus, are 5 and 2 billion years old, respectively, making them among the oldest open star clusters that amateur astronomers can easily observe.

How Far Away Are They?

The Hyades are the closest open star cluster to us, at a mere 130 light years away. This cluster is so close to us that its angular diameter is too great to fit into the field of view offered by most telescopes. The Pleiades is perhaps the best open star cluster in the sky, but at only 440 light years away, it is still spread out across some 2° of space. Owners of wide-field, short focal length telescopes will get the best from this cluster.

Most of the better open star clusters on the Messier list are between about 2,000 and 5,000 light years away. But amateur astronomers do observe some open clusters that are farther away than that. NGC 2158 in Gemini is a good example, a whopping

16,000 light years away. Unsurprisingly, this object is also among the fainter open star clusters and difficult to see with anything less than a 200 mm aperture telescope.

What Can You See Through a Telescope?

Open star clusters are fun! Many of them have received interesting names because of the patterns of stars (or asterisms) detectable within them. Among the more amusingly named open clusters are the Klingon Battlecruiser Cluster (NGC 1662), the 37 Cluster (NGC 2169), the ET Cluster (NGC 457), and no less than two Christmas Tree Clusters (NGC 2264 and NGC 581).

Others are worth observing simply because they are beautiful objects. Examples of showpiece open clusters include the Pleiades, the Praesepe, the Perseus Double Cluster, and the Wild Duck Cluster (M11) in Scutum.

Many open clusters contain stars of different colors as well as double or triple stars. Seeming chains of stars are a common feature. A few are associated with patches of nebulosity, either reflection or emission nebulae, depending on the cluster. All in all these are rewarding objects, and understandably popular among backyard astronomers.

What's the Best Way to View Them?

Since open star clusters tend to be relatively wide in terms of angular diameter, low to medium magnifications are most useful. Light pollution filters aren't usually of much help because they dim starlight, but they may be useful when trying to detect any nebulosity associated with the star cluster.

As with emission nebulae, these large objects are often viewed with low magnification eyepieces. When selecting an eyepiece, it's important to consider the exit pupil, and to choose an eyepiece that doesn't deliver an exit pupil greater than the pupil diameter of a fully dark-adapted eye, i.e., anything greater than 7 mm for young observers, or 5 mm for observers older than 30.

Globular Clusters

What Are They?

Globular clusters are dense swarms of stars tightly bound together by gravity into a distinctive spherical shape. Unlike open clusters, which contain young stars, globular clusters contain very ancient stars. In fact globular clusters are among the most ancient objects in the night sky. Because globular clusters are mostly concentrated around the galactic core, they are observed most between late spring and early autumn, when the constellations of Sagittarius and Scorpio are high in the sky.

How Big Are They?

Globular clusters typically contain several hundred thousand to over a million stars. In terms of size they range from about 60 to 150 light years in diameter.

Given the number of stars packed into such a relatively small volume, the distances between constituent stars are not very great. The Sun, for example, occupies a patch of space with a density of about 0.02 stars per cubic light year, but in the core of a globular cluster stars may be a hundred times more densely packed, with two or more per cubic light year.

If the Sun were a member of a globular cluster, nighttime wouldn't be particularly dark, since the sky would be filled with thousands of very bright stars. Depending of course on how close the Sun were to core, the overall brightness of these stars would be more or less comparable to that of the full Moon here on Earth.

How Bright Are They?

Globular clusters tend to be moderately to very dim objects, and compared with open clusters, relatively few of them are good targets for small aperture telescopes. There are some exceptions though, including a few just visible to the naked eye.

How Old Are They?

As mentioned already, globular clusters are very ancient objects. In fact globular clusters are often studied by cosmologists investigating the age and origins of the universe.

Globular clusters are distinctive in containing mainly Population II stars. These are stars that are notably poor in metals, unlike Population I stars (like our Sun) that are comparatively metal-rich. Immediately after the Big Bang, the universe didn't contain any metals at all, only hydrogen and helium. Consequently the first generation of stars, called Population III stars, contained no metals at all. Like all stars, these Population III stars eventually used up their nuclear fuel and died, and there are none of them left today.

For various reasons these Population III stars were relatively short lived, perhaps lasting for no more than a few million years. But before they died, they carried out nuclear fusion reactions that synthesized new elements, including metals. This is something all stars do towards the end of their lives. Population III stars that became supernovae flung out these newly synthesized elements across space, and some of this material ended up in the nebulae that went on to form the next generation of stars, Population II stars.

Like Population III stars, these Population II stars synthesized metals inside their cores shortly before they died, so that when they exploded as supernovae, they provided even more metals for subsequent generations of stars, known as Population I stars. At this point some qualification of the term 'metal-rich' needs to be made.

Although the Sun contains more metal than the average Population II star, metals still account for just 1.8% of its mass!

Population II stars outnumber Population I stars in our galaxy, but they're mostly concentrated in the galactic core and in globular clusters. Population I stars are mostly found in the spiral arms, and they're of special significance to us because they're the ones most likely to have planets orbiting them.

Because Population II stars were the second generation of stars to form, and did so within a few million years of the Big Bang, they can help pin down the age of the universe. At its oldest, the universe would only be a few million years older than the oldest Population II stars. The coolest white dwarf stars in globular clusters seem to be between 12 and 16 billion years old. Interestingly, although this is broadly comparable to the age of the universe determined using other methods, it isn't precisely the same.

Quite why some globular clusters seem to be older than the age of the universe is a mystery, and a hot topic of discussion among astronomers. One explanation is that the distances between us and the globular clusters has been systematically underestimated, which means that their constituent stars would have to be intrinsically brighter than had been previously assumed. This, in turn, would mean that such stars are hotter and younger, and the age of globular clusters would need to be reduced somewhat, one estimate suggesting by about 15%.

How Far Away Are They?

Amateur astronomers mostly view globular clusters associated with the Milky Way. There are at least 150 such globular clusters, mostly about 200 light years from the center of the galaxy and arranged in a spherical pattern around the galactic core.

Since the Sun is quite a distance away from the galactic core, globular clusters tend to be quite remote. Even the best of them are distant objects when compared with, for example, open clusters. NGC 5139, better known as Omega Centauri, is one of the nearest globular clusters despite being more than 16,000 light years away, while the finest globular cluster in the northern sky, the Great Globular Cluster M13 in Hercules, is still a good 25,000 light years away. Most are even more distant than this.

A few globular clusters, notably M54 in Sagittarius, M79 in Lepus, and NGC 2419 in Lynx, are situated far from the galactic center. In the case of NGC 2419, this globular cluster is 300,000 light years from the galactic core, and about the same distance away from us, making it by far the most remote globular cluster visible through medium to large aperture telescopes.

These globular clusters are or were associated with neighboring galaxies. In some cases they are still orbiting their parent galaxy (as is the case with M54), while others have now been picked up by the Milky Way following some interaction between our home galaxy and one of its satellite galaxies.

What Can You See Through a Telescope?

So far as visual observing goes, globular clusters are both pretty and difficult. They're pretty because of their shape, and under the right conditions they really can look like swarms of stars. Amateur astronomers with large aperture telescopes have often likened them to snowballs or clumps of diamond dust, but observers using small aperture telescopes may be a bit disappointed. Unlike open clusters or planetary nebulae, these are distant, relatively faint objects. The truth is that globular clusters really do reward aperture, and even a 200 mm aperture telescope won't deliver anything like what you see in a photograph.

That said, with averted vision especially, the better globular clusters are unquestionably worthwhile. Observers should be able to detect the bright core of the cluster as well as the fainter, more sparsely populated halo of stars orbiting the core. It should also be possible to resolve at least the outermost stars, and in some cases globular clusters can be resolved much more completely than that. Although most are spherical, not all of them are, and some are distinctly elliptical in shape.

Globular clusters have been classified according to how concentrated they appear when examined through a telescope. Known as the Shapley–Sawyer classification system, the scale runs from I to XII, with I being the most concentrated and XII the least. In practice, globular clusters between V and VII tend to look the best, being concentrated enough that they look impressive, while not so concentrated that they can't be resolved (to some degree, at least) through an average-sized telescope.

What's the Best Way to View Them?

Globular clusters look best under medium to high magnifications. This is partly because it helps to darken the sky, improving the contrast between the stars and the background. But increasing magnification also helps to make it easier to resolve the constituent stars.

The optimal magnification will depend on the aperture of the telescope as well as the size of the globular cluster being observed, but something between ×100 and ×250 is typically best. Of course, increasing magnification reduces overall brightness, which is why averted vision is so important when looking at these objects. Looked at with averted vision globular clusters can often seem to sparkle, appearing brighter and much more resolved than when viewed directly.

Light pollution filters aren't of much use when observing globular clusters. Because they are emitting starlight, narrowband and line filters will simply dim them without providing any benefit. In situations where light pollution is modest, a broadband light pollution might be used, but any improvement to the image will be slight.

Galaxies

What Are They?

Galaxies are huge associations of stars, dust, and dark matter. The stars are, of course, the objects that emit light that we can see, and although dust doesn't emit light, the denser patches of dust, such as dust lanes, can be seen as dark shapes silhouetted against the stars.

The stuff astronomers call dark matter is rather more mysterious. It is thought to account for about 90% of the mass of a galaxy, and although the gravitational effect of all this dark matter is reasonably well supported, dark matter itself hasn't yet been detected. It doesn't emit any radiation, and it doesn't interact with any radiation, which is what makes it 'dark.' Dark matter is important because it lends bulk to the galaxy, and without it the stars in the galaxy wouldn't have sufficient gravity to hold themselves together.

How Big Are They?

Galaxies are, of course, very big indeed. Our home galaxy, the Milky Way Galaxy, appears to be a typical middling to large galaxy. It is about 100,000 light years in diameter and contains up to 400 billion stars.

Substantially bigger galaxies are called giant galaxies. A good example is M87 in Virgo. The main body of this elliptical galaxy is about 120,000 light years in diameter, but it has wispy extensions that stretch out across more than 500,000 light years. M87 contains many times more stars than the Milky Way, likely several trillion stars.

At the other end of the range are the dwarf galaxies. The two Magellanic Clouds that orbit the Milky Way Galaxy are examples of this type. The Large Magellanic Cloud has a diameter of about 14,000 light years, while the Small Magellanic Cloud is only about half that size. Both contain just a few hundred thousand stars.

How Bright Are They?

Galaxies are notoriously dim objects, the classic 'faint fuzzies' that throw suburban astronomers into fits of despair. To be fair, the Andromeda Galaxy M31 is fairly bright, and although it lacks contrast when viewed against light-polluted skies, its bright central core should be visible through even a small aperture telescope. In fact under dark sky conditions M31 is visible to the naked eye.

But M31 aside, and perhaps with the further exception of M33 in Triangulum, northern sky galaxies tend to be challenging objects for observers using small aperture telescopes. Even those galaxies that are considered to be bright by the standards of galaxies, eighth or ninth magnitude, say, are so weak in terms of contrast that picking them out from the background sky can be very difficult indeed. The problem is that galaxies are large objects in terms of angular diameter. They are, of course, intrinsically large objects as well, but the issue here is their angular diameter, because this is an indication of how widely the light they emit is spread out across the sky.

The galaxy M81, for example, has an apparent magnitude of 6.9, which doesn't sound too dim, but its angular diameter of 24 arcseconds means that its surface brightness is very much lower than a magnitude 6.9 star, which is of course a point source of light. In fact under suburban sky conditions the brightest part of this galaxy, its core, is usually the only bit that is easily visible. The surface brightness of M81 is about magnitude 13, and most of the Messier list galaxies have surface brightnesses around magnitude 12–14.

Exurban astronomers have things rather easier, and should be able to pick out the general shape of most galaxies and in some cases dust lanes and spiral arms. But the truth is that if you want to see galaxies properly, you really do need to drag a large aperture telescope out to a genuinely dark sky site.

How Old Are They?

Galaxies are presumed to be nearly as old as the universe, but this isn't to say that they are unchanging.

As noted earlier, the Population II stars that inhabit the core of a galaxy are very ancient, but Population I stars are much younger. So although the Milky Way Galaxy is around 15 billion years old, our Sun, a pretty typical Population I star, is only 4.6 billion years old. The Sun is a small star, and small stars have long lives; giant stars such as Betelgeuse have much shorter lives. Indeed, Betelgeuse is less than 10 million years old but it's life is nearly over and it is all set to become a supernova any time now. When it blows, material it throws out into space may drift into a nebula, triggering star formation, and so another generation of stars will come into being.

Star formation is also triggered by interactions between galaxies. Galaxies move, with clusters of galaxies orbiting one another. Sometimes they drift past one another, and the gravitational pull of one galaxy creates waves to form in the gas and dust of the other. Should the gas and dust clump sufficiently, a nebula can form, and from that nebula are created new stars. Galaxies with unusually high levels of star formation are called starburst galaxies, of which the brightest and most easily seen is probably the Cigar Galaxy M82 in Ursa Major, a galaxy that is being deformed by its much larger neighbor, the giant galaxy M81.

How Far Away Are They?

Galaxies are a long way away. The nearest ones are those that orbit the Milky Way Galaxy, the so-called satellite galaxies.

The two closest satellite galaxies are Canis Major Dwarf Galaxy and the Sagittarius Dwarf Elliptical Galaxy, which are 25,000 and 42,000 light years away from

Earth, respectively. Although no further away than a globular cluster, they're on the opposite side of the galactic core and consequently impossible to see.

A bit further out are the two Magellanic Clouds, the Large Magellanic Cloud about 160,000 light years away and the Small Magellanic Cloud about 200,000 light years away. These satellite galaxies are not visible at all from mid- to far-northern latitudes, but to southern observers these are spectacular objects, easily visible all year round. They are so close to Earth that features such as clusters and emission nebulae within them can be made out by amateurs using even relatively small telescopes.

The Andromeda Galaxy is much further out. This large spiral galaxy is about 2.9 million light years away and 150,000 light years in diameter. Similar to the Milky Way Galaxy in size and shape, like our home galaxy it also has its own attendant group of satellite galaxies, including two that are reasonably easy to spot with medium aperture telescopes, M32 and NGC 205.

Galaxies tend to occur in groups rather than singly. The Andromeda Galaxy, the Milky Way Galaxy, their satellite galaxies, and a number of other galaxies form a congregation known as the Local Group. Among the other groups nearest the Local Group are the M81 Group (11 million light years away), the M101 group (24 million light years away), the M51 Group (31 million light years away), and the Leo Triplet (35 million light years away). A bit further out are the Virgo Cluster (about 59 million light years away), the Ursa Major Group (60 million light years away), and the Fornax Cluster (62 million light years away). Most of these clusters contain at least a few large, relatively bright galaxies that can be seen with medium to large aperture telescopes.

The Virgo Cluster seems to be at the heart of this vast assemblage of galaxy groups, so the whole collection is known as the Virgo Supercluster. This region of space measures something like 110 million light years across, and there's really not much of anything until the next supercluster is reached, the Coma Supercluster about 300 million light years away.

Curiously, galaxies are not evenly distributed throughout the cosmos. Instead they're arranged in streams known as filaments. Superclusters occur where these filaments intersect.

What Can You See Through a Telescope?

Galaxies come in a variety of shapes, and observing the different types is part of what makes them interesting. The two main types are spiral galaxies and elliptical galaxies, but there is also a third catch-all type known as irregular galaxies that don't have any particular shape, or at least can't be classified as either spirals or ellipses.

Spiral galaxies have three main components: the bulge, disc, and halo. The bulge is the spherical core of the galaxy, typically around 20,000 light years in diameter. The disc is a thin circular region that bisects the bulge, at least in part. It is very wide but thin, typically 100,000 light years in diameter but only 1,000 light years in width. Surrounding the entire galaxy is the halo, a region only thinly populated with stars that can be as much as 200,000 light years in diameter. This is the region that contains most of the globular clusters. Population I and Population II stars are not evenly distributed between the regions. The core and the halo mostly contain Population II stars, while the disc mostly contains Population I stars.

The bulge is usually the brightest part of a spiral galaxy, and under suburban skies may well be the only part that is easily visible. Spiral galaxies are of course famous for their spiral arms. There may be two or more spiral arms, often with shorter spurs in between them. Unfortunately for suburban astronomers, spiral arms are notoriously difficult to see under light-polluted conditions. The Andromeda Galaxy is famously frustrating in this regard; although a good spiral galaxy in terms of shape, it is viewed obliquely, and so its spiral shape is difficult to detect. The Whirlpool Galaxy (M51) is much better in terms of orientation, but being viewed face on its overall surface brightness is rather low, making reasonably dark skies essential if its spiral arms are to be detected.

Not all spiral galaxies have spiral arms, though. Spiral galaxies that have the disc but not the arms are known as lenticular galaxies, because they resemble lentils in shape. The Sextans Spindle Galaxy (NGC 3115) is a reasonably bright example of a lenticular galaxy. A further subclass of spiral galaxy is the barred spiral galaxy. These have arms, but instead of spiraling out from the core, these instead merge with a bar-like structure that runs through the core. M58 in Virgo is a good example of a barred spiral galaxy, as is our own Milky Way Galaxy.

The second type of galaxy is the elliptical galaxy. These have bulges and haloes much like spiral galaxies but lack the disc component. So they tend to look more or less spherical, both in photographs and through a telescope. Some are almost perfect spheres, as in the case of M105 in Leo, while others are much more elongated, as is M59 in Virgo.

An interesting observation made by astronomers is that dwarf elliptical galaxies are commonly satellites of large spiral galaxies. The classic examples are the two dwarf ellipticals (M32 and NGC 205) that attend the Andromeda Galaxy. Amateur astronomers should be able to see both of these small galaxies even under slightly light-polluted skies. The Milky Way Galaxy is another spiral galaxy with a dwarf elliptical attendant, in this case the Sagittarius Dwarf Elliptical Galaxy.

The final type of galaxy is the irregular galaxy. These have shapes unlike either spiral or elliptical galaxies, typically more or less cloud-like. The two Magellanic Cloud galaxies that orbit the Milky Way Galaxy are the classic examples, but there are lots of others, though they tend to be fairly dim. The brightest and most easily observed by northern hemisphere observers include the Box Galaxy (NGC 4449) in Canes Venatici and the Cigar Galaxy (M82) in Ursa Major.

In some cases irregular galaxies may well have been either spirals or ellipses, but at some point in time were distorted by the gravity of another, larger galaxy that interacted with them. In the case of M82, this galaxy was deformed from its spiral shape by its much larger neighbor, M81. Several pairs of interacting galaxies are described in this book, among the most famous of which are the Hockey Stick Galaxies, NGC 4656 and NGC 4657, in Canes Venatici; the Siamese Twin Galaxies, NGC 4567 and NGC 4568, in Virgo; and the Antennae Galaxies, NGC 4038 and NGC 4039, in Corvus.

Besides their shape, amateur astronomers will also want to look for structures known as dust lanes. Dust is present in all galaxies, but only in some galaxies, typically large spiral galaxies, does it form dense clouds that block light. The dark nebulae observable in the Milky Way are examples of the types of dust clouds responsible for dust lanes. When galaxies are observed from the outside, the dust will sometimes be seen as streaks around the edges of the galaxy. Several galaxies have dust lanes that can be seen using medium to large aperture telescopes. The classic example is the Black Eye Galaxy (M64) in Coma Berenices, but other galaxies with prominent dust lanes include M108 and NGC 2841, both of which are in Ursa Major, as well as M61 and M104, both in Virgo.

What's the Best Way to View Them?

This is an easy question to answer: the best way to view galaxies is through a large aperture telescope and under truly dark skies! Of course this isn't a very satisfying answer for backyard astronomers lumbered with varying degrees of light pollution, but it is an honest answer.

Under very mild light pollution, a broadband filter may help somewhat, but because galaxies are emitting starlight, narrowband and line filters are of no value at all. If you don't have a broadband filter, raising the magnification to moderate levels (around ×100) can help by heightening contrast while minimizing skyglow, though this trick only works up to a point and won't do anything to help against moderate to severe light pollution. Not all galaxies look good at medium magnifications, and low power, wide-field eyepieces can often be very useful, especially when pairs of interacting galaxies are being observed.

To get the best from your observing sessions, be sure to schedule galaxies for the darkest nights of the year, i.e., when the Moon isn't above the horizon. Be sure to spend at least half an hour observing other deep sky objects before you turn your attention to the galaxies. This will give your eyes time to acquire full dark adaptation and so be most sensitive to dim light. Once you have the galaxy in the eyepiece, try using averted vision. This is often the best way to see elusive features such as spiral arms. Indeed, some galaxies may be impossible to detect at all without using averted vision.

Although it's perhaps a negative way to end this book, there's no escaping the fact that galaxies simply aren't good targets for suburban astronomers.

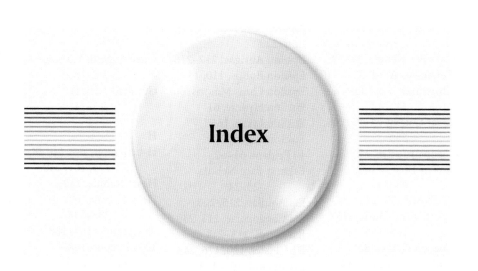

Index